Innovation in Socio-Cultural Context

Innovation—the process of obtaining, understanding, applying, transforming, managing and transferring knowledge—is a result of human collaboration, but it has become an increasingly complex process, with a growing number of interacting parties involved. Lack of innovation is not necessarily caused by lack of technology or lack of will to innovate, but often by social and cultural forces that jeopardize the cognitive processes and prevent potential innovation. This book focuses on the rule of social capital in the process of innovation: the social networks and the norms; values and attitudes (such as trust) of the actors; social capital as both bonding and bridging links between actors; and social capital as a feature at all spatial levels, from the single inventor to the transnational corporation. Contributors from a wide variety of countries and disciplines explore the cultural framework of innovation through empirics, case studies and examination of conceptual and methodological dilemmas.

Routledge Advances in Sociology

For a complete list of titles in this series, please visit www.routledge.com

53 **Globalization and Transformations of Social Inequality**
 Edited by Ulrike Schuerkens

54 **Twentieth Century Music and the Question of Modernity**
 Eduardo De La Fuente

55 **The American Surfer**
 Radical Culture and Capitalism
 Kristin Lawler

56 **Religion and Social Problems**
 Edited by Titus Hjelm

57 **Play, Creativity, and Social Movements**
 If I Can't Dance, It's Not My Revolution
 Benjamin Shepard

58 **Undocumented Workers' Transitions**
 Legal Status, Migration, and Work in Europe
 Sonia McKay, Eugenia Markova and Anna Paraskevopoulou

59 **The Marketing of War in the Age of Neo-Militarism**
 Edited by Kostas Gouliamos and Christos Kassimeris

60 **Neoliberalism and the Global Restructuring of Knowledge and Education**
 Steven C. Ward

61 **Social Theory in Contemporary Asia**
 Ann Brooks

62 **Foundations of Critical Media and Information Studies**
 Christian Fuchs

63 **A Companion to Life Course Studies**
 The Social and Historical Context of the British Birth Cohort Studies
 Michael Wadsworth and John Bynner

64 **Understanding Russianness**
 Risto Alapuro, Arto Mustajoki and Pekka Pesonen

65 **Understanding Religious Ritual**
 Theoretical Approaches and Innovations
 John Hoffmann

66 **Online Gaming in Context**
 The Social and Cultural Significance of Online Games
 Garry Crawford, Victoria K. Gosling and Ben Light

67 **Contested Citizenship in East Asia**
 Developmental Politics, National Unity, and Globalization
 Kyung-Sup Chang and Bryan S. Turner

68 **Agency without Actors?**
New Approaches to Collective Action
Edited by Jan-Hendrik Passoth, Birgit Peuker and Michael Schillmeier

69 **The Neighborhood in the Internet**
Design Research Projects in Community Informatics
John M. Carroll

70 **Managing Overflow in Affluent Societies**
Edited by Barbara Czarniawska and Orvar Löfgren

71 **Refugee Women**
Beyond Gender versus Culture
Leah Bassel

72 **Socioeconomic Outcomes of the Global Financial Crisis**
Theoretical Discussion and Empirical Case Studies
Edited by Ulrike Schuerkens

73 **Migration in the 21st Century**
Political Economy and Ethnography
Edited by Pauline Gardiner Barber and Winnie Lem

74 **Ulrich Beck**
An Introduction to the Theory of Second Modernity and the Risk Society
Mads P. Sørensen and Allan Christiansen

75 **The International Recording Industries**
Edited by Lee Marshall

76 **Ethnographic Research in the Construction Industry**
Edited by Sarah Pink, Dylan Tutt and Andrew Dainty

77 **Routledge Companion to Contemporary Japanese Social Theory**
From Individualization to Globalization in Japan Today
Edited by Anthony Elliott, Masataka Katagiri and Atsushi Sawai

78 **Immigrant Adaptation in Multi-Ethnic Societies**
Canada, Taiwan, and the United States
Edited by Eric Fong, Lan-Hung Nora Chiang and Nancy Denton

79 **Cultural Capital, Identity, and Social Mobility**
The Life Course of Working-Class University Graduates
Mick Matthys

80 **Speaking for Animals**
Animal Autobiographical Writing
Edited by Margo DeMello

81 **Healthy Aging in Sociocultural Context**
Edited by Andrew E. Scharlach and Kazumi Hoshino

82 **Touring Poverty**
Bianca Freire-Medeiros

83 **Life Course Perspectives on Military Service**
Edited by Janet M. Wilmoth and Andrew S. London

84 **Innovation in Socio-Cultural Context**
Edited by Frane Adam and Hans Westlund

Innovation in Socio-Cultural Context

Edited by Frane Adam
and Hans Westlund

NEW YORK LONDON

First published 2013
by Routledge
711 Third Avenue, New York, NY 10017

Simultaneously published in the UK
by Routledge
2 Park Square, Milton Park, Abingdon, Oxfordshire OX14 4RN

First issued in paperback 2014

*Routledge is an imprint of the Taylor and Francis Group,
an informa business*

© 2013 Taylor & Francis

The right of Frane Adam and Hans Westlund to be identified as the authors of the editorial material, and of the authors for their individual chapters, has been asserted in accordance with sections 77 and 78 of the Copyright, Designs and Patents Act 1988.

All rights reserved. No part of this book may be reprinted or reproduced or utilised in any form or by any electronic, mechanical, or other means, now known or hereafter invented, including photocopying and recording, or in any information storage or retrieval system, without permission in writing from the publishers.

Trademark Notice: Product or corporate names may be trademarks or registered trademarks, and are used only for identification and explanation without intent to infringe.

Library of Congress Cataloging-in-Publication Data
 Innovation in socio-cultural context / edited by Frane Adam and Hans Westlund.
  p. cm. — (Routledge advances in sociology ; v. 84)
 Includes bibliographical references and index.
 1. Technological innovation—Economic aspects. 2. Technological innovation—Social aspects. 3. Diffusion of innovations—Economic aspects. 4. Diffusion of innovations—Social aspects. I. Adam, Frane. II. Westlund, Hans.
 HC79.T4I54659 2013
 303.48'3—dc23
 2012030928

ISBN 978-0-415-63622-3 (hbk)
ISBN 978-1-138-92071-2 (pbk)
ISBN 978-0-203-08554-7 (ebk)

Typeset in Sabon
by IBT Global.

Contents

List of Figures	ix
List of Tables	xi
Preface	xiii

Introduction: The Meaning and Importance of Socio-Cultural Context for Innovation Performance   1
FRANE ADAM AND HANS WESTLUND

PART I
The Cultural and Cognitive Framework of Innovation

1 Culture and Cognition: The Foundations of Innovation in Modern Societies   25
MARIAN ADOLF, JASON L. MAST, AND NICO STEHR

2 Socially Responsible Science and Innovation in Converging Technologies   40
TONI PUSTOVRH

3 Culture Impact on Innovation: Econometric Analysis of European Countries   57
PEDRO FERREIRA, ELVIRA VIEIRA, AND ISABEL NEIRA

4 Communities as Spaces of Innovation   80
JOANNE ROBERTS

PART II
Innovation and Social Capital: A Reconsideration of Conceptual and Methodological Dilemmas

5 National and Regional Innovation Capacity through the Lens of Social Capital: A Qualitative Meta-Analysis of Recent Empirical Studies   103
FRANE ADAM

Contents

6 Collaboration in Innovation Systems and the Significance of Social Capital 126
HANS WESTLUND AND YUHENG LI

7 Learning in Regional Networks: The Role of Social Capital 142
ROEL RUTTEN AND DESSY IRAWATI

PART III
Case Studies

8 Independent Inventors and Their Position in the National/Regional Innovation System: A Case Study of Slovenia 161
ANGELA IVANČIČ, DARKA PODMENIK, ALJA ADAM, AND ANA HAFNER

9 "Individuals" Networks and Regional Renewal: A Case Study of Social Dynamics and Innovation in Twente 185
PAUL BENNEWORTH AND ROEL RUTTEN

Discussion and Conclusion

Instead of a Conclusion: Society, Culture, and Innovation: Themes for Future Studies 211
FRANE ADAM AND HANS WESTLUND

Contributors 215
Index 217

Figures

3.1	R&D expenditure per employee, 1995–2009 (100 = countries' average).	69
3.2	R&D expenditure per employee (2009) vs. individualism (index measure).	70
3.3	R&D expenditure per employee (2004) vs. masculinity (index measure).	71
3.4	R&D expenditure per employee (2009) vs. Uncertainty Avoidance Index (index measure).	72
3.5	R&D expenditure per employee (2009) vs. Power Distance Index (index measure).	72
3.6	Cultural dimensions: Portugal, Spain, and European countries' average.	75
6.1	The linear (left) and the interactive (right) models for development of innovations.	130
9.1	The location of the region of Twente in the Netherlands and Europe.	191

Tables

2.1	Worldwide Patenting Trends in Nanotechnology, Biotechnology, and ICT at the EPO and USPTO	43
3.1	Trompenaars & Hampden-Turner (1998) Cultural Dimensions	59
3.2	Hofstede Cultural Dimensions (1983; 1997)	61
3.3	Summary of Research on Innovation and Hofstede's Cultural Dimensions	64
3.4	Variables Definition	68
3.5	R&D Expenditure Econometric Model	73
3.6	Redundant Fixed Effects Tests	73
4.1	The Characteristics of Communities of Practice	85
4.2	Characteristics of Constellations of Practice	86
4.3	Types of Boundary Processes	87
4.4	Varieties of Knowing in Action	90
6.1	Various Dimensions of Collaboration in Innovation Systems with Examples of Variables	137
9.1	The Interviews within the Case Study by Time and Sector	192

Preface

This is a book that from the point of view of common sense should not be needed. It should be obvious for everyone that innovation—something new and creative—is not just a product of technology and capital on the micro level, or business life, research and government on macro level. It should be so obvious that innovation takes place in a social context of human interaction and cultural values. Still, when we started our planning of this book, we found a surprisingly limited literature and even less empirical evidence on this topic. Thus, we hope that the book will start to fill a gap that is much wider than we originally thought.

We would not have been able to write this book by ourselves, but if this book is seen as an innovation, it is very much a product of our social context. We were lucky to find a number of high-level scholars that wanted to contribute to the book. The manuscript was read and commented on by Professor Roger Bolton, Williams College, who also came up with valuable suggestions for the final chapter. Ivana Čančar, IRSA, has done a great job as our editorial assistant. The Slovenian book agency (JAK) provided funds for language editing for Slovenian authors.

To all who have contributed to this book and to our colleagues at IRSA, we express our heartfelt thanks.

Ljubljana and Stockholm, July 2012
Frane Adam and Hans Westlund

Introduction
The Meaning and Importance of Socio-Cultural Context for Innovation Performance

Frane Adam and Hans Westlund

INTRODUCTION

Innovation and a New Combination of Productive Knowledge

A new agenda for the social sciences is called for today because the age of labour and (economic) capital is at an end as the sole driver and paradigm of economic productivity. Work itself is not at an end, but the nature and volume of the labour required are rapidly changing. The traditional forces of production are less and less the motor of economic development and value-adding activities. Nor are the conventional productive means the most important source of any sustainable economic activity. We are witnessing the growing significance of knowledge and other intangible forms of capital.[1] (Post)modern economies are ever more innovation-driven economies. However, it cannot be forgotten that the recent economic crisis is (also) a consequence of the uncontrolled expansion of global financial markets. Financial capital remains—with all contradictions an important determinant of global and national economies.

The transformations of the economy, as a result of the rising importance of knowledge, go far beyond the mere emergence of the 'knowledge company,' the 'knowledge creating firm,' or the firm 'as a system of knowing activity.' The issue now concerns the model of sustainable socio-economic development in modern knowledge-based societies (Moldaschl and Stehr, 2010). One of the most challenging issues is how to connect sustainability and basic solidarity with the notion of meritocratic society. This type of society namely creates real incentives for individual achievements, innovativeness, and creativity.

Democratic knowledge-based society can be conceived of as involving the coincidence of two processes, namely, the scientification of society and the socialisation of science (and technology). Furthermore, innovation activity can hardly be valued merely for its technological component in terms of new products, patents, and innovation processes, but should also be valued for its organisational, cultural, and civil society aspects.

The scientification of society connotes a process where science and its technological applications penetrate all societal segments. Scientific findings and their applications, innovations, information (and ICT infrastructure), knowledge, and human capital (including so-called tacit knowledge) are becoming the main force driving production and reproduction (identity) in (post)modern societies. The phenomenon of new emergent sciences and technologies like bio-, nano-, info-, and cognitive sciences, which in the framework of mutual interconnectedness form new complexes of transdisciplinary convergent technologies, is opening up a whole new dimension of social and economic development. We might expect that, because of certain fundamental findings and related innovations in the abovementioned scientific fields, already in the near future the social and economic performance of knowledge-based societies will be evaluated according to completely new criteria. On the other hand, since scientification and technology are placed within a (civil) society framework every scientific solution and application demands a very well considered approach, including from the perspective of its placement in macro- and micro-social environments, and in view of its long-term consequences and risks (see Nowotny, Scott, and Gibbons, 2001; Pustovrh, Chapter 2, this volume).

The Triple Helix concept that demonstrates new relationships between science and other subsystems of society can thus be translated as the interpenetration of industry, academic science, and the state (government). One outcome of this new constellation is the emergence of academic entrepreneurship as an important driver of hi-tech innovations (Etzkowitz, 2011). This conceptualisation also implies interpenetration zones in the sense of the supporting or intermediary agencies and institutions that have been formed within this multilateral conglomerate.

Cognitive mobilisation is a broader concept than innovative capacity since it is not reducible to the commercialisation of knowledge and research. Similar terms are knowledge mobilisation or the "cogitisation of society" (Etzkowitz, 2003) and to some extent also knowledge ability (see Chapter 1 by Adolf, Mast and Stehr, this volume). Cognitive mobilisation denotes the increasing exposure of individuals, organisations, regions, and nations (as well as supra-national entities like the EU) to "scientification" processes, meaning the emergent significance of scientific logic including in everyday life ("Lebenswelt"), the rapid acquisition of new information and knowledge-sharing as well as interactive learning (see Adam et al., 2005). This process is accompanied or in some cases corrected by the parallel process of the socialisation of science or its social contextualisation. Cognitive mobilisation refers to the capability not only for action in the sense of applying and utilising scientific and technological knowledge, but also for a critical reflection on the risks of 'scientification' (and its technological applications). This capability renders long-term strategic policy on science and technology (S&T) possible, harnessing it in the framework of sustainable development.

In contrast, national innovative capacity is "the ability of a country to produce and commercialize a flow of innovative technology over the long term" (Furman, Porter, and Stern, 2002). The results of innovative capacity are patent applications regarding new products and services, trademarks, industrial design, organisational and marketing innovations, and innovation processes (Innovation Union Scoreboard, 2011). Cognitive mobilisation is important for innovative capacity, yet it may result in an increasing knowledge stock, part of which is not intended or anticipated for marketable purposes but which contains ingredients enabling *strategic, long-term thinking and interactive (lifelong) learning* (Lundvall, 2006) as well as the *capability for (self-)reflection*. All these components of cognitive mobilisation which enhance learning capacity and creativity on the level of society accelerate and improve innovation performance. Namely, they enable creative re-combinations of different types of knowledge.

Joseph Schumpeter (1912) stated that innovations are new combinations of production factors. In the knowledge economy of today, innovation is largely about new combinations of the production factor *knowledge*. New combinations happen through spontaneous spillovers but also via organised transfers of knowledge. New knowledge is created from new combinations, reflected in the fact that the first part of Schumpeter's claim, "new combinations," is today often discussed in terms of *creativity* (also see Trigilia, 2002). Among other things, this creativity is expressed in the ability to identify the different types of knowledge[2] to combine and to perform the combinations. It can be argued that the presence of social capital is an input in the production of human capital and knowledge and that it facilitates the creative combination of the production factor of knowledge. According to another explanation, social capital may even be viewed as a production factor or a component of total factor productivity (Dasgupta, 2001).

SOCIETAL CONTEXT, INNOVATION, AND SOCIAL CAPITAL

This book focuses on social, institutional, and cultural factors which to a lesser or greater degree and directly or indirectly exert an influence on innovation processes and innovative performance on different levels.

Our research and thematic platform is organised around the issues of the connection and contribution of social capital and related concepts like networking, trust, cooperative capability, organisational culture, and value orientations to innovation intensity as well as knowledge transfer and the creation of innovative milieus.

Proceeding from this frame of reference and by taking an overview of recent publications into account, it could be said that our book will introduce new dimensions to the discussion on the innovation-driven knowledge economy (and knowledge-based society). It seems that it will be one of the very rare books so far to have thematised the socio-cultural framework in such an integral way.

Innovation capacity and innovativeness in general can be studied and interpreted from different points of view. An increasingly accepted thesis is that developmental dynamics and the innovation process (which are closely linked to creativity, knowledge transfer, and knowledge management along with entrepreneurial initiative) cannot be fully understood without reference to social and cultural contexts. However, until now relatively little empirical evidence has been available regarding which components of this context are relevant and how they facilitate or hamper this process. Our book has the intention and ambition to contribute to a more concrete analysis and more accurate interpretation.

One of the strong accents in the book is the role of social capital research and application, and the impact of the related concepts of interpersonal trust, networks, and collaborative structures on innovation performance by taking account of different levels—the cross-national, national, regional, and firm—as well as individual inventors.

Bonding and Bridging Social Capital

A traditional perspective generally sees knowledge as something exogenous an innovator collects and combines, whereas creativity is an endogenous, inherent quality of an inventor. However, this view must be called into question. Schumpeter already pointed out that negative reactions of the social environment would prevent innovation. Alfred Marshall (1880) highlighted the positive sides of the social environment when he wrote about the "air" of Britain's 19th century industrial districts and the knowledge spillovers that occurred there. Thus, there is a good basis to argue that the social environment has an impact on innovation or, in other words, that *innovation takes place in a socio-cultural context.*

What then is this socio-cultural context? While a single common definition does not exist, it seems as if the concept is often used synonymously with the concepts of social environment, social context and milieu (see, e.g., Wikipedia). As in the case of the abovementioned concepts, there is no common, uniform definition of social capital across (or even within) disciplines. By defining social capital here as social networks and the norms and values distributed in these networks, we put ourselves in the mainstream of the social capital literature. However, we reject the notion that social capital always is a 'good' and that poor development can be explained by 'too little' social capital. As shown by, e.g., Portes (1998), there are many examples of destructive social networks, norms, and values. Accordingly, depending on its qualities, social capital can exert a positive or a negative influence on innovation.

There are many theories about innovation, its component parts, and its diffusion in time, in space, in the economy, etc. One of the best known theories of innovation is Rogers's (1962/1995) model of diffusion of innovation over time, according to which there are five stages of this process:

1. Knowledge: The recipient is informed that the innovation exists and about its function.
2. Persuasion: The recipient becomes persuaded of the value of the innovation.
3. Decision: The recipient decides to embrace the innovation.
4. Implementation: The recipient puts the innovation to use and starts exploiting it.
5. Confirmation: The recipient uses the innovation in full (or takes it out of use).

Analogously with the discussion in Westlund's and Li's chapter in this book on the importance of social capital for collaboration in innovation systems, it can be stated that social capital plays a role in all five of Rogers's stages: Both the knowledge and the persuasion stages are characterised by communication between the agent of the innovation and the recipient—a communication in which social networks, values, and attitudes can play a decisive role. The innovation agent's ambition is to establish social capital that is as 'bonding' as possible with the recipient, while the recipient is initially looking for alternative solutions, thereby keeping up 'bridging' social capital with links to many alternative agents. The decision to embrace the innovation is a result of establishing bonding social capital with one of the agents of the innovation. When it comes to implementing the innovation, it is other parts of the recipient's social capital that become important, viz., the relations with employees and other collaborators, and with the early group of customers who are the most willing to test and adopt innovations.[3] The social capital of this fourth stage is also partly bonding, not only to the innovator but to the actors involved in the implementation process. In the confirmation stage, the recipient has conquered a share of the market by, among other things, building brands and other social relations with the anonymous mass of customers (cf. Westlund 2006, Chap. 5). This is done by creating new, bridging links to new groups of customers and, in the successful cases, transforming these bridging links into links with bonding features, through, e.g., customer clubs, goodwill, and other 'emotional' relations.

It must be kept in mind that Rogers's model was developed in the manufacturing industry's golden years and that it focuses on marketing and commercialisation when an innovation (or invention) already exists. Yet in today's knowledge economy the situation is different. The focus of both research and policy is increasingly being put on the creation and emergence of innovation. Thus, there are reasons to focus on earlier stages of the innovation process, the stages where creativity and knowledge interact and develop new knowledge. What role is social capital playing in the interaction of creativity and knowledge?

In contrast to the era of the 'lonely geniuses' behind innovations of the 19th century, most of today's innovations are the result of teamwork and extensive collaboration between a large number of actors (Westlund, 2009). It would of course be a gross exaggeration to say that the lonely geniuses existed in a vacuum without any social relations, but it is self-evident that current knowledge production depends on a much more comprehensive and extended social capital.

Social Capital inside and outside Firms

An important feature of the definition[4] of social capital is that it distinguishes between the networks and the norms, values, and preferences that are distributed in the networks. The dual role of social capital—in certain circumstances promoting spatial and group internal cohesion, in other circumstances contributing to link-building that promotes spatial interaction across physical and social borders—makes social capital a much more complicated factor than the trust-building, transaction-cost-reducing factor often assumed in the modern literature. In the spatial perspective, certain component parts of social capital work by internalising and are governed by the actors in that locality/region, and other component parts give access to externalities which local/regional actors cannot govern. A similar conclusion can be drawn from the firm's perspective. The internal social capital of a firm is formed by its management and employees, and for its survival there are strong incentives to internalise firm-specific knowledge—but via location decisions and other investments the firm also builds links and relations to gain access to externalities such as knowledge and information.

Following Johansson (2004), we can assume that knowledge transfers and collective learning take place through two types of processes:

1. deliberate, formalised transaction-links, agreements, networks, and other club-like arrangements between firms and firms and other actors; and
2. unintended knowledge spillovers between firms or between firms and other actors, caused by non-formalised interactions. These kinds of interactions consist of: (a) vertical technical/economic interactions between firms and their suppliers and/or customers; (b) spin-offs of new firms from existing ones and turnover and exchange in the labour market; (c) horizontal interaction in the form of the informal exchange of information and knowledge in the (local/regional) civil society, between individuals connected to firms or other actors.

In the first case, that of formalised transaction-links and networks, the formalisation is itself a confirmation of the firm's willingness to invest in a link with a longer duration than a pure market transaction. In contrast to the 'conventional wisdom' on (spatial) clusters, the reasons for the emergence of these fixed links/networks are not the firm's wish to enjoy the informal knowledge and information spillovers and other outcomes of flexible inter-actor interactions—but to *internalise* knowledge within the fixed network, often a corporate grouping. McCann and Arita claim that the cluster type of Silicon Valley is more an exception which should not be generalised, and that the internalising "industrial complex" type of cluster is "typical of many firms and sectors, and in particular, of the semiconductor industry" (McCann & Arita 2004, 247). In this case, knowledge spillovers within a delimited industrial complex are internalities for the industrial complex, but still extra-market externalities for the individual firms—externalities that are both formalised and institutionalised.

It can be assumed that the motives for a link-investment are completely based on economic considerations, yet the outcome of this 'long-term' investment depends among other things on the social relations between the actors who establish, use, and maintain the link. With negative attitudes to the link among these actors, incentives to use the link would be lower, and the link would yield lower returns than in the case of neutral or positive attitudes. Thus, it is in the interest of all the actors who invest in the link or network to establish positive social capital among its users.

Civil Society and Contextual (Regional) Social Capital[5]

Civil society is basically something that is formed and maintained by people during their non-productive time. Voluntary public and club activities and other leisure activities are also what civil society's social capital is focused upon. The networks and values of business life, i.e., of production, play a mainly hidden role in civil society. In line with the fundamental differences between production and consumption, business life and civil society are based on different principles and belong to different spheres of human activities, with different networks and different norms and values. However, as the two spheres are populated to quite a considerable extent by the same people, i.e., the productive population, there are naturally certain informal interactions between them. These interactions can be divided into two types: (a) those mainly based on norms, values, attitudes, etc.; and (b) those where these values have developed into the links and networks of a group of individuals.

The first type of interaction includes general approaches concerning the importance of 'spirit' and similar attributes to the economic development of a region. Putnam's (1993, 2000) view of the impact of civic society on the economy and Florida et al.'s (2002) view of creativity as a factor constituting an important difference between a region's economic performance are examples of such spatially connected approaches.[6] It is reasonable to assume that the attitudes of the social environment that have an impact on the production environment in general are formed in the interplay of the two spheres. A stable production sphere fosters stable attitudes in civil society and vice versa. The old industrial regions of the world provide many examples of how this stability of business life and civil life created safe and predictable conditions for stable growth. On the other hand, when the industrial crisis came in the 1970s these regions lacked the ability to change and, in Schumpeter's words, "to do something new."

The other type of interaction in civil society with implications for business life is a result of the general values of the social environment, in which the values of communities, groups, or sub-communities have developed into links and networks. While a community can in principle be based merely on some kind of shared values, the step toward the formation of informal networks of groups or sub-communities means more stable relations between certain actors in the community—and a way to partly avoid the reactions of general opinion. Spatial agglomeration as such provides a potential solution to the problem of gaining from the change-promoting elements of social capital and avoiding the restricting elements. This potential solution builds on the fact that agglomerations tend to foster the emergence of diverse groups and sub-communities, based on ethnicity, religion, industry, and also interests. Florida et al. (2002) and other scholars examining creativity stress the importance of this tolerated diversity in a limited space and regard the interfaces between these groups as an important source of creativity, innovations, and economic development.

Within a group or sub-community that views innovations positively, an entrepreneur can find the support he or she needs "to do something new." Outside this sub-community, the entrepreneur can pick up certain information and knowledge of other groups, where this knowledge is not too tacit and internalised. In this way, the agglomeration can provide the entrepreneur with access to the social capital that supports his or her activities the most and provides the best access to useful tacit knowledge.

It can be said that civil society is an important source and generator of social capital, which may also be beneficial for business activities on the level of firms. Put differently: it is here we encounter the crossing of different circles (or, in Simmel's words, "die Kreuzung der sozialen Kreise") and networks. However, this does not mean these two spheres are in a harmonious relationship, for it is often characterised by conflict and distrust. It must be underlined that entrepreneurs and the rest of the population differ with regard to some values and norms (Beugelsdijk, 2010).

Some investigations suggest the crucial role of regional/civic social capital regarding innovation success on the level of firms (Masciarelli, 2011; also see the overview in Adam's contribution in this volume, Chap. 5). However, so far little research has been done on the crisscrossing and intertwining of networks belonging to the different spheres of society. Proceeding from recent empirical research, proposals are made for the location of ambitious firms: "the results of this chapter suggest that contextual settings in terms of social capital should be taken into account by managers and entrepreneurs planning their innovation strategies. Firms might deliberately choose to locate in geographical areas with high levels of social capital" (Masciarelli, 2011, 33). Unfortunately, the research design and presentation of findings in Masciarelli's book do not allow such settings/regions to be identified. The approach taken is variable-based and (sophisticated) statistical calculations are made on an aggregate level (at the level of the entire sampling of regions and on the basis of mean scores). The reader may be frustrated since there is no indication at all of which specific (Italian) regions are better off regarding their levels of social capital or creativity. Without a comparative analysis encompassing other regions from other countries (given that such datasets are readily available—like Eurostat or the European Innovation Scoreboard—now called the Innovation Union Scoreboard), it is also very difficult to assess innovation performance or the real strength of social capital, and the author recognises this (ibid., 77).[7]

In addition to the study of the impact of (contextual) social capital on the innovation success of firms measured by the introduction of new product or technological process is the impact of regional creativity (measured by human capital including R&D employees as well as cultural capital in the form of the population's participation in cultural events). This—otherwise competently written—book includes also the significance of entrepreneurial social capital/networks (actually Masciarelli is focused on managers) for innovation (measured by the position generator technique) has also been investigated. In both cases there are positive correlations. Yet it seems that geographically bound (regional) social capital plays a superior role in the sense of moderating and mediating between these both variables and innovation (ibid, 29–30; 76).

Social Networks without Social Capital?

While several authors have recently dealt with innovation performance by concentrating on the term social capital, some use this term very rarely or do not use it at all, even though they consider similar phenomena and processes. In the volume *Innovation in Complex Systems* (Ahrweiler, ed., 2010), a number of authors refer to social networks, innovation networks, collaborative arrangement, without mentioning or using the term social capital. One reason for omitting it might be the intention of the authors (or editors) to employ more exact and hard 'scientific' instruments. They call for the combination of a quantitative social science method with natural science approaches. It is argued that the innovation process can be modelled in a mathematical way and be made predictable, despite the claim that this process is not linear but highly complex and even chaotic. It is believed that "Agent-based modeling of innovation processes can advise actors of innovation networks in a way which makes innovation indeed computable . . . Developing computational designs for innovation policy laboratories *in silico* can help to inform policymakers and business managers about optimal network structures for collaborative research and innovation" (Ahrweiler, ed., 2010, 320). It is both interesting and revealing that this author draws attention to the significance of qualitative case studies that "provide the required in-depth knowledge" (ibid., 15). Can this be interpreted as recognition that sophisticated quantitative tools are insufficient to acquire new insights and cognition? Can this claim pave the way for 'soft' factors like social capital or values?

In another influential book about social networking and innovation (where the term social capital also does not appear), the author argues that network analysis is the best method for capturing the "essence" of innovativeness (Liebowitz, 2007). However, he points out the pitfalls of a pure quantitative network approach and recommends a triangulation strategy by saying: "Interviews and focus groups also delve deeper into the areas that the survey highlighted" (ibid., 38). It is important that he stresses the significance of culture that allows new ideas and stimulates knowledge-sharing.

CULTURAL CONTEXT OF INNOVATION

Since the end of the 1980s we have witnessed the revival of the culturalist approach (often associated with historical dimensions) employed by social scientists from different schools of thought and research fields: path-dependency, neo-institutionalism (historical institutionalism), modernisation theory, intercultural communication, and organisational culture theory. This 'cultural turn' (Sztompka, 1999) has also proved to be potentially productive in expanding the conceptual repertoire of developmental paradigms. One should mention here at least three representative approaches. Berger coined the term economic culture in order to show the importance of a broader matrix within which economic processes take place (Berger, 1987). Putnam introduced the notions of civic culture and civic tradition (connected with the concept of social capital) as factors strongly influencing institutional and economic performance (Putnam, 1993). Inglehart systematically investigates the issue of changing value orientations, i.e., the shift from materialist to post-materialist values and its (cor)relations with economic growth and the modernisation profile of a given nation or region (Inglehart, 1997). In addition, some economists have contributed valuable insights in this regard (North, 1990; Greif, 1994; Tomer, 2002), and the same is true of certain more ethnographically oriented analyses of cultural diversity and the distinctiveness of national or regional cultural patterns in business and intercultural communication (Trompenaars, 1993; Lewis, 2000).

Culture in this context is generally conceived as a certain *forma mentis*, a symbolic-semantic fund, which, against the background, informs and orients individual and collective behaviour. It is a kind of institutional software, a 'mental programme' (Hofstede, 1980). While there are many definitions of culture, we would like to point out the following: "The culture is the invisible force behind the tangibles and observables in any organisation, a social energy that moves people to act. Culture is to the organisation what the personality is to individuum—a hidden yet unifying theme that provides meaning, direction and mobilisation" (Killman, 1985, cited after Carayannis and Campbell, 2011, 4). Whether culture (values)—among other factors—shapes the ways people think about and behave with regard to their work and profession, risk taking, time perspectives, cooperation, and competitiveness, its influence on economic activity as well as the propensity for technological and social innovations can be assumed or taken for granted.

However, the significance or range of its influence can only be understood in interaction with structural factors and social actors' decisions, in terms of the configuration of subjective and structural points of view (DiMaggio, 1994). In this context, we note the following statement by Granovetter: "More sophisticated . . . analyses of cultural influences make it clear that culture is not a once-for-all given, but an ongoing process which is continuously constructed and reconstructed during interaction. It not only shapes its members but also is shaped by them in part for their own strategic reasons" (Granovetter, 1985, 486).

It may be said that the inherited cultural models are the subject of trial-and-error modifications actors make in order to cope with contingencies emanating from the structural environment (Patterson, 2000, 210). Expressed in other words, culture is a 'tool kit' providing resources for facilitating or limiting an actor's capacity to construct an appropriate action strategy (Swidler, 1986). We illustrate this point in the following section by referring to the structural environment which forced actors to abandon their traditional passive role and to redefine the value system in order to legitimate their more effective collective action. In sum, we must not fall into cultural (historical) determinism, nor should we relativise cultural factors in the belief that changes in structural and formal rules inevitably lead to changes in customs and values compatible with the new institutional structure (see North, 1990).

Generally speaking, one cannot understand culture as either an independent or a dependent variable, but more as an intervening variable (see Hungtington & Harrison, 2000). Or, as Berger (1987) argues, here one does not presume culture in terms of values, beliefs, and lifestyles to determine what occurs in the economic sphere, or vice versa; establishing a (possible) cause-effect relationship that requires empirical analysis and case studies. This is also a general point of departure for our discussion of historically derived cultural dispositions, which are thought to contribute to the mobility and 'availability' of human resources in terms of increasing productivity and achievement motivation in the context of the long-term modernisation process. Referring to Sztompka (1999) and his concept of civilisational competence, we must avoid possible misunderstandings when we define a certain country, region, or community as more or less civilisationally competent. Here we can rely on Parsons who argued that the increase in adaptive capacity (which is, however, connected with the value generalisation process) as the main characteristic of modernisation (civilisational competence making significant contributions in this regard) does not "imply moral superiority" (Parsons, 1977, 70).

Civilisational competence can thus be understood as a psycho-cultural and socialisation pattern, which has been sedimented and transmitted from generation to generation, may be accumulated, and in certain circumstances is open to innovative change (invention of tradition). It is a latent structure of cognitive, normative, expressive, and motivational elements which enables individuals and social communities to orient themselves in the different subsystems of modern (or modernising) societies (Adam et al., 2005).

Regarding the impact and strength of culture and values, different interpretations can be seen and therefore two questions are important. First, are cultural patterns changeable and flexible or are they more stable and persistent over a longer time? Second, it is also unclear whether they can be considered as a determinant (independent) variable or merely as a residual explanatory factor. However, these two issues are connected: authors who consider culture as flexible are not inclined to attach a deterministic significance to it. In contrast, authors who stress the stability insist on its path-dependent character. One of these authors is Hofstede and the same is true of Sztompka. The former is a very influential author and his concept of cultural orientations has been used by many researchers dealing with entrepreneurship and innovation. On the other side, Inglehart with his cultural zones (and post-materialist orientation) represents a different position. It can basically be understood in the sense that economic development and modernisation trigger cultural changes that make individual autonomy, gender equality, and democracy increasingly likely. Value changes reflect changes in economic and political life, but with a time lag. However, the cultural shift also has an impact on economic and political development and cultural factors contain considerable autonomy and strength of their own (Inglehart, 1997).

In contrast with this dynamic approach to cultural change, Hofstede argues that "National values systems should be considered given facts, as hard as a country's geographical position or its weather" (Hofstede et al., 2010, 21). One author recently applied one of Hofstede's dimensions (uncertainty avoidance) to the propensity for entrepreneurial activity by saying that "Cultural aspects are time invariant" (Suddle, Beugelsdijk and Wennekers, 2010). The problem of this genuine deterministic (path-dependent) approach is that it is quite an *a priori* proposition for which there is no empirical evidence in the sense of reliable longitudinal or panel survey data. The main empirical evidence Hofstede (and other authors) relies on largely stems from research around the 1970s. It is interesting that this author and other authors (see Freytag and Thurik, eds., 2010) still use this 40-year-old data with an (unproven) argument about the persistence and stability of national values on the basis of a principle which could be formulated as follows: "values do not change easily (or at all) and we can use the old data without hesitation"!

Besides the very recent monograph *Culture and Entrepreneurship* (Freytag and Thurik, eds., 2010) where the authors chiefly deal with the impact of culture on entrepreneurial activity and partly regional innovativeness, several research reports focusing on the significance of cultural patterns for innovation performance have been published (Kaasa and Vadi, 2010; Williams and McGuire, 2010; Vechi and Brenan, 2008; Ferreira et al., Chap. 3, this volume). They employ Hofstede's categories and, with the exception of Kaasa and Vadi (2010), also his (old) dataset. It may be concluded that their findings are relatively incongruent or inconclusive. Most of them reach agreement (similar results) in just one category, namely, uncertainty avoidance (UA). However, Fereira et al. (Chap. 3, this volume) came to the conclusion that UA does not have a significant impact on innovation on the European level, but when taking Spain and Portugal separately it can be seen that high levels of uncertainty avoidance hamper innovation performance as measured by patent applications. Concerning other categories (power distance, individualism-collectivism, masculinity-femininity—a long-term orientation is very rarely utilised), the results are even more divergent.

It is very difficult to synthesise these studies even though the authors use the same categorical apparatus borrowed from Hofstede and even his dataset, since they perform different sampling which is surely a reason for the differing results, while another one is probably the outdated dataset (Ferreira et al., Chap. 3, this volume). With those authors who use other indicators/items from a different dataset (mainly from EVs and ESS), is unclear if their indicators/items are comparable with those of Hofstede (Kaasa and Vadi, 2010; Beugelsdijk, 2010). However, it is quite obvious that also in this research field more deliberation, dialogue, and coordination are needed to create more consistent and consensual findings and interpretations concerning the impact of cultural values on economic and innovation performance. In contrast, it can be seen—and this is fascinating—that the notion of a 'cultural turn' which emerged at the intersection of sociology and cultural studies has found resonance in all social science disciplines (and beyond). The same is true of other socio-cultural dimensions, especially social capital. Yet it seems that these dimensions cannot be fully captured and interpreted by conventional "positivist/empiricist" methods; therefore an epistemic and methodological shift is needed in order to provide a more appropriate and complex research strategy.

THE CONTRIBUTIONS OF THIS BOOK

This book addresses and discusses most of the issues elaborated in the first part of the introduction. It can be said that it is one of the rare publications which is thought to present and evaluate (most) recent research findings regarding the socio-cultural context of the innovation process. Although the authors come from various social science disciplines, their common denominator is the impact of social capital and cultural values on this process. Some of contributions are empirically based, some are written rather in the form of theoretical essays, but all are engaged in seeking new answers to the question how to utilise knowledge and innovativeness for the steering of societal complexity on micro and macro levels.

Part 1 The Cultural and Cognitive Framework

The contribution written by sociologists Marian Adolf, Jason L. Mast, and Nico Stehr (Chapter 1) is a (meta)theoretically motivated chapter. These authors investigate the social and cultural conditions that foster innovativeness in the context of modern, knowledge-intensive economic systems. In contradistinction to the dominant legal, institutional, and economic approaches, they offer a social theoretical approach based on an understanding of the major trajectories of modern society, which seeks to identify the cultural and cognitive prerequisites of innovative thought and creativity. They also comment on the difficulties that defining innovation poses for social theory and frame innovation as cognitive displacement, whereby existing metaphorical frameworks are used to discover and account for new phenomena in a process that changes the composition of both the metaphor and the new phenomenon. It is suggested that knowledgeability, or a bundle of social competencies, emerges as the main prerequisite for the potential of innovative thinking in a society that is characterised by fragility and uncertainty.

The next contribution, Socially Responsible Science and Innovation in Converging Technologies, by Toni Pustovrh, a political scientist dealing with social studies of science and technology, is based on a broader collection of data and an analysis of relevant publications. The chapter examines several societal mechanisms intended to foster 'socially robust' S&T by strengthening social capital within and among the main actors in the S&T production process. Building on developmental trends in the new technologies expected to serve as future economic drivers, it examines mechanisms such as codes of conduct and civil society organisations which could promote socially beneficial and desired innovations, while increasing trust and reciprocity. In concluding, it affirms the need to strengthen social capital in various societal segments involved in the S&T production process as an important element of modern societal cohesion and resilience.

The chapter titled Culture Impact on Innovation, by the economists Pedro Ferreira, Isabel Neira, and Elvira Vieira, is a quantitative empirical study. The starting point is that innovation is made up of a complex set of processes strongly relating to contextual factors. In this sense, the cultural environment is crucial for countries to be innovative. This chapter contributes to understanding the effects culture has on innovation and tests an econometric model when considering the relationship between Hofstede's Cultural

Dimensions Theory and innovation. The results show that three out of four cultural dimensions do have an impact on innovation. These results hold practical implications, namely, the fact that some countries can present more innovative potential than others and, consequently, are in a better position to be competitive and develop entrepreneurial activities. Based on this framework, the cases of Portugal and Spain are examined in more detail and thus complement the variable-based approach with a case-based approach.

Joanne Roberts's chapter titled Communities as Spaces of Innovation is a theoretical work based on a synthetic overview of the literature on communities of practice. A community-based perspective complements established approaches to understanding and promoting the creation of knowledge and innovation in spatially specific contexts whether at an urban, regional, or national level. These approaches include innovation systems, clusters, industrial, innovative milieus, and networks. Many of these approaches focus on the macro dimensions of innovation such as institutional and policy structures. In contrast, a focus on community offers the potential to explore the social interactions that underpin knowledge creation and innovation at a micro level. Roberts begins by exploring the turn to community in contemporary economic thought before exploring early conceptualisations of communities of practice. She then turns her attention to the size and spatial reach of communities of practice and the need to differentiate between communities of practice of different activities, along with the types of knowledge, spatial reach, duration, and nature of social ties. The challenges and opportunities of communities of practice as spaces of innovation are then outlined. The chapter concludes with a brief assessment of the place of communities of practice among approaches to understanding innovation at various spatial scales together with tentative policy suggestions to promote the role of community as a facilitator of innovation.

Part 2 Innovation and Social Capital: A Reconsideration of Conceptual and Methodological Dilemmas

Chapter 5, National and Regional Innovation Capacity through the Lens of Social Capital, by the sociologist Frane Adam, is an empirical study that aims to reconstruct the methodological design and findings of 17 mostly quantitative investigations of the impact of various dimensions of social capital on both national and regional innovation capacity. This reconstruction of empirical findings paints a relatively ambiguous picture. It seems that only a radical modification of research designs and methods can lend new impetus and allow us to gain a more in-depth understanding of the role of social capital and/or its dimensions in (regional) innovation processes. Regarding the existing body of knowledge, the meta-analysis approach is appropriate for discerning the most relevant, grounded, and comparable theses and data. Yet, in the case of new research, more attention should be paid to the ethnographic approach, case studies, and triangulation.

Chapter 6, by Hans Westlund and Yuheng Li, Collaboration in Innovation Systems and the Significance of Social Capital, takes its starting point in the abovementioned fact that innovation occurs in systems of actors, i.e., innovation systems.[8] The collaboration between participating actors is crucial to the success of innovation. Many forms of routine collaboration and exchange of goods and products work well without any real need to take care of social relations. However, if the question arises of the exchange and collaboration of human resources such as knowledge—and not least in terms of combining existing knowledge with new knowledge—then issues of *social capital* may play a critical role in their success or failure. Collaboration between companies, research bodies, governmental agencies, and other actors in an innovation system can take an almost infinite variety of different forms. Westlund and Li present a typology of the different dimensions of collaboration and its component variables. They show that social capital is important for the performance and outcomes of a large number of collaboration dimensions. It is concluded that innovation policies should pay greater attention to the social capital of innovation systems. However, the authors underscore the need for further studies before detailed policies can be developed: meta-studies of how actors' social networks and values influence collaboration in innovation systems; meta-studies of existing studies, and evaluations of innovation programmes and projects to analyse the potential of policies to influence social capital among actors in innovation systems. Finally, they also suggest the shaping of a European research programme on data collection concerning collaboration in innovation systems.

Chapter 7, by Roel Rutten and Dessy Irawati, Learning in Regional Networks, deals with the question of why some regions perform better in the global economy than others, or Why do some regional networks perform better than others? To answer this question, the authors focus on one hand on the characteristics of regional networks and, on the other, on the characteristics of the region where a network is embedded. Both of these problems have been addressed by recent literature, but Rutten and Irawati argue that one crucial element remains underdeveloped: social capital. This chapter makes the case for the role of social capital by identifying how it plays a function with respect to learning in regional networks. The authors start by explaining the role of social capital with regard to learning in networks. The second section presents a short case study of the Western Java automotive industry. This case study is relevant as social capital plays a different role in this industry than it does in similar networks in regions in the developed world. The third section takes a closer look at the spatial dimension of networks and social capital. Recognising that social capital is also found at the regional level, the fourth section elaborates on how these two levels—(networks of) individuals and regions—can be conceptually connected with regard to the role of social capital in relation to learning. Since social capital is involved in the social interactions between individuals and because learning is a social process among individuals, the level of individuals in regional networks seems to be the most appropriate level for studying the role of social capital in regional learning processes. The final section presents a summary and conclusion of the arguments made in this chapter and suggests that an 'individual turn' is needed in economic geography to explain social capital's role with regard to learning in networks.

Part 3 Case Studies

Chapter 8, titled Independent Inventors and their Position in the National/ Regional Innovation System, by Angela Ivančič, Darka Podmenik, Alja Adam, and Ana Hafner, is a qualitative study focusing on the importance of social capital for independent inventors in overcoming barriers deriving from their marginalised position in the innovation system and from the absence of organisational resources enjoyed by corporate inventors. Taking into consideration the literature arguing that innovation and technology are gendered areas, this issue was also explored from the gender perspective. Based on a case study and results of an analysis of 22 interviews (plus one focus group) with independent inventors in Slovenia, it investigates main sources of their social capital and the ways they collaborate with the external innovation supportive environment. The structural approach and the individual approach to social capital are used for the explanation. Social capital is seen as an individual investment in social networking which allows access to different sorts of resources and can contribute to different sorts of profits.

The findings suggest that a lack of opportunities as well as the capacities for collaboration in open networks, especially formal ones, is crucial for access to institutional resources. The low level of trust in people and institutions represents an important barrier to co-operation of independent inventors in teams and their participation in various networks. Civil society structures and peer organizations are pointed out as the main sources of the structural as well as individual social capital.

The findings also confirm that gender plays an important role in access to social capital and external support. It appears that closed networks provide more support to male inventors than to female inventors. However, in terms of establishing relationships to external networks, both men and women are equally hindered by their distrust in institutions.

Chapter 9, Paul Benneworth's and Roel Rutten's Individuals' Network and Regional Renewal, addresses the research question of how the social habits, norms, and routines of local innovation networks shape the wider culture and behaviour of their particular local territory, and how they influence the local innovation environment as well as economic development outcomes. To discuss this question, they begin by looking more closely at the concepts of Territorial Innovation Models (TIMs), and highlight the continuing critique of their failure to analytically specify social variables and dynamics as more than an emergent function of success. They then explore a case study of a region in which there has been a shift in its learning capacity which, while not representing a totalising cultural shift, still represents a qualitative change in its culture associated with—but not entirely driven by—a qualitative change in its innovative capacity. This process of cultural change is analysed by focussing on the way social dynamics on a variety of scales—the individual, the network, the regional-political, and the regional-cultural—interact as part of this change. These different scales are linked by hinging activities which bring, for example, individuals together within networks. What TIM exponents consider to be change in the regional culture is built upon parallel but separate changes in individuals' networks as well as innovation transaction networks. Therefore, the authors conclude that a more comprehensive approach to studying regional learning cultures needs to make this very distinction significantly clearer to restore rigour to the study and deal more meticulously with the interrelationship between these scales at which innovation takes place.

NOTES

1. It is known that the notion of human capital was first introduced to (neo-classical) economics by T.W. Schultz and G. Becker (who also dealt with social capital). It is still worth quoting a short statement Schultz made in 1962: "Truly the most distinctive feature of our economic system is the growth in human capital" (cited in Moldaschl and Stehr, 2010, 29).
2. There are different forms of knowledge (besides codified and tacit)—in most cases, this notion means expert or science-derived knowledge. One author (Asheim, 2010) distinguishes innovations built on analytical knowledge (scientific knowledge used in strong R&D based hi-tech industries), innovations built on synthetic knowledge (more practical technical, or engineering knowledge used in a mature sector in non-R&D-based economies), and innovations built on symbolic knowledge (arts-based knowledge used in cultural creative industries).
3. Rogers also divided the customers into five groups according to their willingness to adopt innovations.
4. "Social capital is considered as a type of infrastructure with nodes and links. The nodes consist of individuals and organizations, which establish links between each other. The construction of links is governed by the individuals'/organizations' norms, preferences and attitudes, which can thus prevent emergence of links between individuals or organizations as well. In the links, different types of information are distributed between the nodes. Social capital's impact on society depends on both its quality and quantity. The norms, preferences and attitudes of the nodes, and thereby the kind of information being distributed in the links, is at least as important as is the number of links. A 'strong' social capital can thus have preservative as well as progressive effects, depending on its qualitative characteristics" (Westlund, 2004, 10).
5. Parts of this section are based on Westlund (2006).
6. Weber's (1904/05) essay on the Protestant ethic is a classic example of this approach. Even if Weber focused on differences in norms and values because of religion, the cases he chose gave his essay a spatial dimension.
7. The author recognises this problem by saying: "Future research should analyse the relationship between social capital and innovation using longitudinal

data," as well as: "Future research should test the hypotheses in this chapter using a sample on firms located in other countries, which would be informative about different country partners" (Masciarelli, 2011, 77).
8. It should be noted that here 'innovation system' should be interpreted literally and thus has a broader meaning than the policy concept of innovation system.

REFERENCES

Adam, F., Makarovič, M., Rončević, B., & Tomšič, M. (2005) *The challenges of sustained development: The role of socio-cultural factors in East-Central Europe.* Budapest, New York: Central European University Press.
Ahrweiler, P. (ed.). (2010) *Innovation in complex social systems.* London: Routledge.
Asheim, B. (2010) Innovation is small: SMEs as knowledge explorer and exploiters. In P. Ahrweiler (ed.), *Innovation in complex social systems.* London: Routledge.
Barnett, E., & Casper, M. (2001) A definition of "Social Environment." *American Journal of Public Health,* 91 (3), 465.
Berger, P. (1987) *The capitalist revolution.* Aldershot: Wildwood House.
Beugelsdijk, S. (2010) Entrepreneurial culture, tegional innovativess and economic growth. In A. Freytag & R. Thurik (eds.), *Entrepreneurship and culture.* Berlin, Heidelberg: Springer Verlag.
Carayannis, E.G., & Campbell, D.F.J. (2011) *Mode 3 knowledge production in quadruple helix innovation systems: 21st-century democracy, innovation, and entrepreneurship for development.* (SpringerBriefs in Business). Berlin, Heidelberg: Springer Verlag.
Dasqupta, P. (2001) *Social capital and economic performance—Analytics.* Available at *http://www.eui.eu/Personal/Guiso/Courses/Lecture%206/Dasgupta_social_capital.pdf.* Accessed January, 2012.
DiMaggio, P. (1994) Culture and economy. In *The handbook of economic sociology,* ed. Neil Smelser & Richard Swedber. New York: Russel Sage Foundation.
Etzkowitz, H. (2003) Innovation in innovation: The Triple Helix of university-industry-government relations. *Social Science Information,* 2003, 42 (3), 293–337.
Etzkowitz, H. (2011) Normative change in science and the birth of the Triple Helix. *Social science Information,* 50 (3–4), 549–568.
Florida, R., Cushing, R., & Gates, G. (2002) When social capital stifles innovation. *Harvard Business Review.* Available at: *http://hbr.org/2002/08/when-social-capital-stifles-innovation/ar/1.* Accessed May 2010.
Freytag, A., & and Thurik, R. (eds.). (2010) *Entrepreneuership and culture.* Berlin, Heidelberg: Springer Verlag.
Furman, J.L., Porter, M.E., & and Stern, S. (2002) The determinants of national innovative capacity. *Research Policy,* 31, 899–933.
Granovetter, M. (1985) Economic action and social structure: The problem of embededness. *American Journal of Sociology,* 9 (3).
Greif, A. (1994) Cultural beliefs and the organisation of society: A historical and theoretical reflection on collectivist and individualist societies. *Journal of Political Economy,* 102 (5), 912–950.
Huntington, S., & Harrison, L. (eds.). (2000) *Culture matters: How values shape human progress.* New York: Basic Books.
Hofstede, G. (1980) *Culture's Consequences: International differences in work related values.* Beverly Hill, CA, Sage.

Hofstede, G., Hofstede, G.J., & Minkov, M. (2010) *Cultures and organizations: Software of the mind (3rd ed)*. New York: McGrawHill.
Hofstede, G., Noorderhaven, N., Thurik, A., Uhlaner, L., Wennekers, A., & Wildeman, R. (2004) Culture's role in entrepreneurship: Self-employment out of dissatisfaction. In T. Brown & J. Ulijn, *Innovation, entrepreunership and culture. The interaction between technology, progress and economic growth*. Cheltenham and Northampton: Edward Elgar.
Inglehart, R. (1997) *Modernization and postmodernization:Cultural, economic, and political change in 43 societies*. Princeton: Princeton University Press.
Innovation Union Scoreboard (2011) Research and innovation union scoreboard. Available at http://ec.europa.eu/enterprise/policies/innovation/files/ius-2011_en.pdf Accessed February 2012.
Johansson, B. (2004) Parsing the menagerie of agglomeration and network externalities. CESIS Electronic Working Paper Series, Paper No. 2. www.infra.kth.se/cesis/research/ workpap.htm. Accessed December 12, 2005.
Kaasa, A., & Vadi M. (2010) How does culture contributes to innovation? Evidence from European countries. *Economics of Innovation and New Technology*, 19 (7), 583–604.
Kahin, B. (ed.). *Advancing knowledge and the knowledge economy*. Cambridge: Harvard University Press.
Lewis, R. (2000) *When cultures collide*. London: Nicholas Brealey Publishing.
Liebowitz, J. (2007) *Social networking: The essence of innovation*. Lanham, MD: Scarecrow Press.
Lundvall, B.-Å. (2006) *Interactive learning, social capital and economic performance*. In Foray, D. and Kahin, B. (eds.), *Advancing Knowledge and the Knowledge Economy*, Harvard University Press, US.
Marshall, A. (1880/1920) *Principles of economics: An introductory volume*, 8th edition, 1920. London: Macmillan.
Masciarelli, F. (2011) *The strategic value of social capital*. Cheltenham: Edward Elgar.
McCann, P., & Arita, T. (2004) Industrial Clusters and regional Development: A Transaction-Costs Perspective on the Semiconductor Industry. In: de Groot, H.L.F., Nijkamp, P., Stough, R. (Eds.) *Entrepreneurship and Regional Economic Development: A Spatial Perspective*. Cheltenham: Edward Elgar, 225–251.
Moldaschl, M., & Stehr, N. (eds.). (2010) Eine kurze Geschichte der Wissensoekonomie. In Moldaschl, M., and Stehr, N. (eds.), *Wissensoekonomie und Innovation*. Bonn: Metropolis-Verlag.
Moldaschl, M,. and Stehr, N. (2010): *Wissensekoenomie und Innovation*. Marburg: Metropolis-Verlag.
North, D. (1990) *Institutions, institutional change and economic performance*. New York: Cambridge University Press.
Nowotny, H., Scott, P., & Gibbons, M. (2001) *Re-thinking science: Knowledge and the public in an age of uncertainty*. Cambridge: Polity Press.
Parsons, T. (1977) The Evolution of Societies. New Jersey: Prentice Hall.
Patterson, O. (2000) Taking culture seriously: A framework and an Afro American illustration. In *Culture matters*, Lawrence E. Harrison & Samuel P. Huntington (eds.). New York: Basic Books, 202–219.
Portes, A. (1998) Social capital: Its origins and applications in modern sociology. *Annual Review of Sociology*, 24, 1–24.
Putnam, R.D. (1993) *Making democracy work*. (Civic Traditions in Modern Italy). Princeton: Princeton University Press.
Putnam, R D. (2000) *Bowling alone: The collapse and revival of American community*. New York: Simon & Schuster.

Rogers, E.M. (1962/1995) *Diffusion of innovations.* New York: Free Press.
Schumpeter, J.A. (1912/1934) *Theorie der wirtschaftligen Entwicklung.* Leipzig: Duncker & Humblot. English translation published in 1934 as *The Theory of Economic Development.*
Swidler, A. (1986) Culture in action: Symbols and strategies. *American Sociological Review,* 51, 237–286.
Suddle K., Beugelsdijk, S. and Wennekers, S. (2010) Entrepreneurial Culture and Its Effect on the Rate of Nascent Entrepreneurship, in: Freytag A. and Thurik R. (eds), Entrepreneurship and Culture, Heidelberg: Springer.
Sztompka, P. (1999) *Trust.* Cambridge: Cambridge University Press.
Tomer, J.F. (2002) Intangible factors in the Eastern European transition: A socio-economic analysis. *Post-Communist Economies,* 14 (4), 421–444.
Trigilia, C. (2002) *Economic sociology.* Oxford: Blackwell.
Trompenaars, F. (1993) *Riding waves of culture.* London: Nicholas Brealey Publishing.
Vecchi, A., & Brennan, L. (2008) A cultural perspective on innovation in international manufacturing. *Research in International Business and Finance,* 23(2), 181–192. http://www.sciencedirect.com/science/article/pii/S027553190800024X. Accessed January, 2012.
Weber, M. (1904/05) *Die protestantische Ethik und der "Geist" des Kapitalismus,* Archiv für Sozialwissenschaft und Sozialpolitik 20: 1–54 and 21: 1–110. English translation: *The Protestant Ethic and the Spirit of Capitalism,* translated by Talcott Parsons, 1931/2001, London; Routledge Classics.
Williams, L. K. & McGuire, S. J. (2010) Economic Creativity and innovation implementation; the entreprenerial drivers of growth—Evidence from 63 countries, Small Bus. Econ., 34, 391–412.
Westlund, H. (2004) *Social capital and the emergence of the knowledge society: A comparison of Sweden, Japan and the USA.* Östersund: ITPS.
Westlund, H. (2006) *Social capital in the knowledge economy: Theory and empirics.* Berlin, Heidelberg, New York: Springer Verlag.
Westlund, H. (2009) The social capital of regional dynamics: A policy perspective. In Karlsson, C., Andersson, Å.E., Chechire, P.C., & and Stough, R.R. (eds.), *New directions in regional economic development.* Berlin, Heidelberg, New York: Springer Verlag, 121–142.
Wikipedia. http://en.wikipedia.org/wiki/Social_environment. Accessed May 2012.

Part I
The Cultural and Cognitive Framework of Innovation

1 Culture and Cognition
The Foundations of Innovation in Modern Societies[1]

Marian Adolf, Jason L. Mast, and Nico Stehr

In *The Wall Street Journal* (Asia; December 10–12, 2010, 12), a deputy principal of the Beijing University High School comments on the recent results of Shanghai's 15-year-olds who topped the global league tables in the PISA tests and notes perhaps the obvious, namely, that "Chinese schools are very good at preparing their students for standardizing tests." However, he adds, "for that reason, they fail to prepare them for higher education and the knowledge economy." Indeed, knowledge-intensive economic systems represent a new economic epoch. The same comment concludes on a cautionary note: China "has no problem producing mid-level accountants, computer programmers and technocrats. But what about the entrepreneurs and innovators needed to run a 21st century global economy?" In this chapter, we will try to answer the question of what competencies and skills are important for innovativeness in a knowledge-based economy. Obviously, we agree, success at responding to clues of standardized testing cannot be the answer.

Although we are stressing the importance of discontinuities, it is not our contention that the sum of the discontinuities constitutes a historically new economic system. According to Werner Sombart (1916–1917, 2122) writing in the tradition of the historical school—every economic system has a form of organization, a technique, and a mental attitude. Of these attributes of the economic system, the unique set of attitudes toward economic life at different times, for example the principles of acquisition, of competition, and of economic rationality of the capitalist system, are among the most important. We are not proposing that the knowledge-intensive economy has such a unique leading idea[3], a leading idea that would allow one to readily identify the historically distinctive mental trait of the knowledge-based economic[4]. Nor can it be said that knowledge-based economic systems have a unique form of organization or technique. Some observers are convinced that modern information and communication technologies represent a distinctive technique typical for knowledge-intensive economies. But as we maintain, the operative economic function of information and communication technologies often tends to be overstated. There is considerable continuity in "economic evolution." Last but not least it is the continuity of the necessity of capital to reproduce itself.

But with Sombart (as well as Max Weber), we are stressing that culture and cognition generally make a considerable difference in economic affairs. The importance of culture and cognition grows as we move toward a knowledge-intensive economic system[5]. This observation about the significance of culture and cognition has another social theoretical merit in that it keeps us from the familiar attempt to conceive of modern society in economic terms only, referring, for example, to the relentless economization of society.

We are claiming that certain cognitive competencies not only tend to be more common in knowledge-based economies but operate as a mental prerequisite for creativity, innovation, innovative processes, and the comparative advantage of nations. We are asking what specific orientations, competencies, and characteristics must a person or collectivity have in order to be innovative or take innovative ideas from his or her environment on board, taking for granted that institutions provide important conditions for the possibility of innovations (cf. Modaschl, 2010, 1).

By stressing certain cognitive competencies as the foundation for innovation, we are not merely reiterating the much more common observation about the growing importance of a highly educated or skilled labor force in modern economies. In fact, we are offering a rival hypothesis, namely, that the cognitive capacities enhancing innovativeness—although associated with formal education and years of education—are not only among the foundations for innovation as we see it but also among the foundations for educational achievement (cf. Schieman and Plickert, 2008, 176). We see our contribution to the innovation literature as a contribution to the sociology of innovation in distinction to the now dominant economic theory of innovation (cf. Godin, 2010). We diverge from the mainstream economic literature, for example, from Giovanni Dosi (1984, 88–89) who, in the field of industrial innovation, sums up the conditions for the possibility of technological innovation in market economies as best described and served by the dual conditions of technological opportunity and the private appropriability of the benefits of the innovative activities. The commitment of private firms to innovation (in contrast to the capability to be innovative) is, of course, undeniably linked to their ability to temporarily appropriate the marginal additions to new knowledge, and therefore the economic advantages that may accrue from the control over novel knowledge (also Geroski, 1995). We are stressing that in addition to these necessary technical, legal, and economic factors, cultural and cognitive prerequisites are the sufficient condition for the possibility of invention and innovation.

After all, as more and more innovation studies have shown[6], the realisation of knowledge, or its translation into technical artefacts, is an extremely complex intellectual and organisational process that relies on sources of knowledge and on "action networks" both "internal" and "external" (for example, on "public science"; see Gibbons and Johnston, 1974) to firms or organizations[7].

The social process of innovation does not follow consistent patterns. In the case of innovation routes, we are dealing with a rather fragile social process, riddled with disappointment, that does not lend itself to exact planning and prognostication (cf. Latour, 1993; Gibbons et al., 1994). As Richard John (1998, 205) shows, in a study of the evolution of American communications, "the most fundamental technical breakthroughs—electric signaling in the 1840s, voice transmission in the 1870s—emerged in highly unusual contexts that provide few obvious lessons for students of innovation today. Equally idiosyncratic was the conceptual advance that hastened the creation of the modern postal system in the years immediately following the adoption of the federal Constitution."

We are going to advance our argument about the importance of culture and cognition for innovation in a number of steps. First, we will address the notion of innovation and argue that we do not have a general theory of innovation, last but not least because the concept of innovation is, as it were, all encompassing. We shall also make the point that the distinction between invention and innovation proposed by Joseph Schumpeter ([1911] 1934) in his 1911 seminal volume *The Theory of Economic Development* is rather difficult to sustain in a consistent manner. Second, in light of the broad notion of innovation as, in the end, of any kind of change, we will focus on the idea that innovation (invention) represents a process of cognitive displacement. Third, we offer some additional observations about the culture of inventions and innovation in societies and in economic systems that are fragile, uncertain, tenuous, and face wicked problems that are most difficult to solve—if they are even capable of being solved. Fourth, for our purposes, we frame innovation as a process of a form of cognitive displacement, whereby existing metaphorical frameworks are used to explain new phenomena in a process that changes both the metaphor's and the new phenomenon's compositions. Fifth, we suggest that the phenomenon knowledgeability, or bundle of social and cognitive competencies, emerges as the main prerequisite for the potential of innovative thinking. We conclude by examining the most important social competencies that structure the possibilities for invention and innovation.

INNOVATION

Few words in any language are as frequently employed as is innovation. Perhaps innovation, on this stage, can compete with democracy and knowledge. All three terms are difficult to define. Nor are there many words that consistently meet with such partiality and approval as innovation. Innovation carries strong normative connotations. If I put it more formally, the term innovation typically performs the speech-act of commending what it tries to describe (cf. Sartori, 1968; Broman, 2002, 5).

It is therefore difficult to separate normative from analytical elements in the case of the concept of innovation. Although the most common conception of innovation refers to the successful implementation of a novel idea, it is complicated (cf. Beckenbach and Daskalakis, 2010), perhaps impossible, to separate the genesis of a new idea (invention) from its practical realization (innovation). Moreover, and as Steve Woolgar (1998, 442) has emphasized, whether or not "ideas count as new, necessarily depends on the social networks involved." A novel idea is not self-validating but has to be recognized as such by other social actors. At the same time, there are few other social phenomena that are generally more significant for modern societies than innovation, knowledge, and democracy. Even political regimes that are authoritarian systems prefer to claim that they constitute innovative societies and have innovative economies.

In one of its most recent publications, the European Commission (2010, 13) offers the following definition of innovation: "whether the innovation is a product, process, marketing method or organizational method it must be new (or significantly improved) to the firm." This definition therefore encompasses not only changes that are entirely novel, but also those that are adopted from outside. The Commission with slight understatement, perhaps even irony, adds: "Our capacity to measure and understand the process in practice still needs work."

It is poignant that the idea of innovation plays such a central role in much of our contemporary political discussion about the economy, the wealth of nations, competitive advantages of societies, but that we appear to be unable to arrest and fence in the notion of innovation itself. The notion of innovation in the sense of novelty is also contained in such concepts as social change, development, evolution, mutation, creation, growth, imitation, invention, modernization, revolution, progress, discovery, and so on. In other words, there cannot possibly be a general theory of innovation since this would amount to a theory of life itself. The concept of innovation refers to processes, namely, change or novelty that is as least as universal as its opposite, namely, routine or habitual conduct. Innovations as such can hardly be predicted: "Genuinely new ideas come out of the blue" (Vromen, 2001, 199).

Hence if one desires to talk sensibly about innovation (invention), one must proceed with a relational concept of innovation[8]. For example, if one wants to account for technical innovations one does not need a theory of technology since technology only evolves in the context of society and not by itself. However, what would be required is a socio-economic theory of technical innovation. Such a theory would refer to a combination of factors such as the creativity of social action, economic incentives—as we mentioned earlier—and institutional conditions (or, on a smaller scale, social or action networks) that enhance technical innovativeness (cf. Moldaschl, 2010, 14).

In the case of Joseph Schumpeter's theory of social change within firms, the yeast that propels change within this set of complex factors is the creative entrepreneur. Our relational concept of innovation concentrates on those features of the subject or the collectivity that enable innovation.

Subjects of course are embedded within a specific social context that either validates a displacement as new or resists such a declaration about its own social network.

CULTURE AND COGNITION: STRUCTURES, TOOLS, PERFORMATIVES, AND METAPHORS

In normative cultural terms, inventors and innovators are cast as geniuses, as social misfits, as loners, and often as psychologically tormented, as social actors tinkering in an unsettled cognitive state that becomes the wellspring for innovative ideas[9]. These cultural frameworks indicate the intangibility of the processes that create and facilitate inventive and innovative thoughts. If they are the product of geniuses, after all, then they are in many ways superhuman, of near-otherworldly origins; if innovators are simply born that way, then they are inexplicable and beyond the grasp of scientific investigation. Innovative ideas just happen to brilliant people, the framework suggests. We must conclude that on this fundamental point the very cultural categories that we bring to bear to understand invention and innovation are precisely part of the cognitive conditions that inhibit innovative and inventive thought, because the normative cultural understandings define innovative thought as indefinable, or as "ah-ha" moments that happen to brilliant people. But for our sake, these cultural forms of understanding make our question even more pressing and interesting: can we shine a light on the conditions that foster and encourage innovative thinking?

The issue of the potential for creative thought lies at the center of cultural-theoretical debate. Culture is theorized as representing a "tool kit," or set of interpretive frameworks that social actors can choose selectively to solve problems and make sense of things during "unsettled cultural periods" (Swidler, 1986, 280), on the one hand, but also as a deep system of structuring symbol systems, a structure that we are born into, with very little capacity to reflect back on (Saussure, 1983 [1916]), on the other. In these opposing theoretical formulations, culture is both handily accessible and usable, such as the words we use, for instance, but it is also a system of understanding that we cannot fully control, such as the rules of grammar and sentence structure. Social actors can choose their words, for instance, but to communicate their ideas, it is words (or another conventional symbol system) that they must use, and the means by which they assemble and organize their words into thoughts are determined by conventions that exist beyond their experience and control. It is this relative autonomy of the cultural system—existing outside any one user's control—that makes communication and understandings between people possible and comprehensible. But this very feature, culture's stable, structure-like character, renders innovation theoretically problematic. We are left with a tension between a structural theory of culture that virtually renders impossible the very notion of innovative thought versus a pragmatic theory that suggests symbols are more like tools that social actors can use strategically and manipulate to solve problems when confronted by unsettling conditions.

The speech act theory of the philosopher of language, John Austin (1975 [1962]), offers a theoretical pathway through this antinomy. Austin brought the issue of innovation to the fore when he identified "performatives" as words that create new social understandings when they are uttered instead of simply making truth claims. Performatives do things, he pointed out, while constatives are merely true or false. Thus, performatives can be understood as structurally available symbols that social actors can invoke to innovate, to suggest new social understandings. When a couple utters the words, "I do" during a wedding ceremony, for instance, they have altered the social landscape, they have created something new, a marriage, and this innovation will dictate that individuals as well as social institutions will treat them in substantially new ways.

Put one way, an innovative concept is by definition a felicitous speech-act: something that once uttered brings the contents of the speech into reality and social being. It is a new formulation that when uttered forces us to see things in a new light. Judging a concept innovative then is an ex-post facto practice: brilliant ideas are uttered all of the time, yet very few of them make it into the popular currency. As Austin pointed out, oftentimes speech acts are infelicitous, or unhappy, in that they fail to create new understandings and social conditions. If part of our project to is to identify the social conditions that facilitate or inhibit innovative thinking, then we are also interested in when a new concept is capable of being felicitous, and the conditions in which it fails to resonate with the broader social arena, and can be interpreted as infelicitous. Why do some innovative concepts succeed and create new understandings, and under what conditions do concepts fail to capture intellectual and popular imaginations? These are questions about the social contexts of reception, a topic that this chapter points to as a topic for future theoretical consideration.

As the structural theory of culture suggests, most social contexts and the vast majority of social action are characterized by nothing more than routines attributes and habitual conduct. The French sociologist and anthropologist Gabriel Tarde introduced the notion that most human action is based on imitation. As an aside, habitual action has of course the constructive function of stabilizing human conduct, enhancing the predictability of social action, and opening up avenues free from the pressures and constraints to act.

Whether or not habitual conduct or imitation is always a carbon copy of previous social conduct (in that limited sense all conduct is a modification of previous conduct) is not at issue; what is at issue is the overwhelming constraint in social life to repeat and therefore get on with life[10]. Nonetheless, within a historical perspective ranging across the centuries, the volume and the speed with which modifications of social conduct occur have of course accelerated with the dawn of the industrial society[11].

Having noted that defining innovation itself is an exercise in trapping a chimera in a bottle, we nonetheless offer a broad definition as an orienting tool: conceptual innovation refers to the epistemological realm between a paradigm shift, on the one hand, and explaining new experiences and phenomena with existing theories or understandings, on the other. Our subject lies somewhere between a radical epistemological break and the steady and routine march of the majority of scientific practice, that of fitting ever more newly flavored wine into old, familiar interpretive bottles. If Austin's theory of the performative offers a cultural theoretical opening for change and innovation, then Donald Schön's ([1963] 1967) work on conceptual displacement offers a bridge between culture and cognition by arguing that it is through metaphorical extension, the application of existing metaphors to new conditions and problems, that innovative ideas may be born.

Building on Schön's ([1963] 1967, 53) investigations into innovative concept formations, conceptual innovations can be described as a process of the displacement of concepts, that is, the "shift of old concepts to new situations," puzzling experiences, or phenomena. The old concept becomes "a symbol or metaphor for the new situation." The new concept then evolves as a result of the work that goes into the making, elaboration, and correction of the metaphor. Cognitive invention or innovation occurs through this process of conceptual displacement as both the old metaphor and the new experience are changed in the process of bringing them together. The metaphor changes, and the new experience is shaped into something more familiar, yet distinct, its peculiar characteristics made intelligible and somehow more mundane. Cognitive displacement refers to the entire working out process or spelling out process of a new metaphor. As Schön ([1963] 1973, 57) points out, the displacement of concepts always occurs in specific contexts from, which as he put it, the source of energy comes. The displacement of concept may be speculative or playful, for example, "as when a child is amused at the idea of a boiling tea-kettle as a baby crying, or a biologist is intrigued with the notion of heredity as the transmission of coded information."

FROM CULTURE TO KNOWLEDGE

One of the significant references in discussions about the role of innovation in the modern economy is knowledge or knowledge production as a foundation stone for innovativeness (cf. Beckenbach and Daskalakis, 2010). If one defines knowledge as a capacity of action, as the ability to set something in motion, invention and innovation constitute the ability to generate novel capacities for action. And if one follows Austin's theory of the performative and Schön's concept of invention as cognitive displacement, then novel capacities for action or knowledge represent noticeable and acknowledged departures from habitual and routine responses to the contingencies of social action.

In this sense, therefore, the extraordinary importance of scientific and technical knowledge does not primarily derive from its peculiar cultural image as representing essentially uncontested (or, objective, that is, reality-congruent) knowledge claims. In this context, the tremendous importance of scientific and technical knowledge in developed societies is related to one unique attribute of such knowledge[12], namely, that it represents incremental capacities for social and economic action or an increase in the ability of "how-to-do-it" that may be "privately appropriated," if only temporarily, in as much as the benefits from innovations based on incremental knowledge are stretched out or leaked to third parties[13].

But the puzzle that remains is: what attributes of individuals and groups enhance their ability to offer non-routine responses to situations in which the habitual responses fail to achieve desired outcomes, for example, the desire to enhance the competitiveness of a firm[14]. We would like to offer the idea that it is knowledgeability that provides the conditions for the potential of invention and innovation. Our conception of knowledgeability, as should become evident, does not merely constitute one of the basic foundations of innovativeness alone. Knowledgeability represents social and cognitive competencies that generally amount to the ability to master one's life more fully.

KNOWLEDGEABLITY

Thus far we have moved from the tension between structural and pragmatic theories of culture, to the cultural-theoretical spaces of innovation embedded in the concepts of the performative and metaphorical displacement. We now introduce knowledgeability to move us further into the realm of the cognitive, and to address the social sites in which innovative actors must bring to bear their cognitive competencies in order to create and protect their creations. Put another way, knowledgeability represents mainly the cognitive prerequisite for agency, or the nexus of cognition and action; we intend for it to constitute the bundle of social and cognitive competencies that actually drive the process of invention and innovation participation.

Our definition of knowledgeability refers neither to what is called common sense, non-reflexive or ordinary knowledge, nor do we refer to specialized scientific-technical knowledge. Also, knowledgeability should not be conflated with knowledge, especially not with its frequent proxy in empirical studies, namely, years of schooling. Knowledgeability is closer to what is at times defined as reflexive or theoretical knowledge. For example, knowledgeability should be seen as the ability of actors and groups of actors to move items of concern onto a particular agenda, such as bottom-up innovationm for example (cf. Von Hippel, 2006)[15].

Unpacking the bundle of competencies means to enumerate some of the important specific cognitive and social capacities conferred by knowledgeability, capacities that are mobilized in accordance with the demands of specific contexts. We list below the most important social and cognitive competencies that drive the possibility of inventions and innovation in modern society:

The capacity to exploit discretion: Since the social rules and legal norms and regulations that govern ordinary and extra-ordinary social conduct are never constituted and enforced in ways that do not allow for discretionary interpretation and execution, the competence to mobilise discretion refers to the capacity of individuals to gain comparative advantages, for example, in such areas as taxation, investment, schooling, and income. The capacity to exploit discretion is essential in efforts to escape the constraints of routinised social action.

The faculty to decide refers to the ability to make decisions and plans and to explore alternative avenues of social action, even in the face of resistance[16].

The facility to organise protection: The capacity to put protective devices and measures in place is a matter of specialised competence that enables actors to mobilise access to differential knowledge in order to ensure, for example, that assets and entitlements are protected against structural or inordinate depreciation. The symbolic or material opportunity costs of the failure to organize protection can be considerable[17].

The authority to speak (cf. Bourdieu, 1975; Lyotard, [1979] 1984) and effectively participate in the workplace or in society extends, for example, to the ability to introduce items on the corporate or political agenda, or to dissent from dominant agendas. The authority to speak in order to dissent applies for instance to many features and situations[18] in everyday life but also extends to the ability of lay audiences or persons to enter a discursive field of expertise as "speakers confront the alleged truth of the discourse that justifies those practices" (Larson, 1990, 37).

The ability to mobilise defiance constitutes another crucial component of the stratifying mode of knowledgeability. To challenge the practices of experts, the state, or corporations constitutes an important asset of knowledgeability as a capacity to contribute to non-routine conduct. In the same sense, the ability to evade surveillance by superiors, the state, or in the marketplace, to formulate discourses of resistance, and to obtain spaces of self-regulated autonomy acquires considerable significance and is based on the capacity to mobilise tools that typically are otherwise seen as instruments exclusively designed to enhance scrutiny.

The ability to access different knowledge bases: The capacity to gain access to different forms of knowledge including those of adjacent fields can be considered to be a crucial resource for setting the process of cognitive displacement in motion. It is of course in this sense that trans- or interdisciplinary training acquires special significance[19].

In general, therefore, the range of social competencies amounts to stratified facilities for mastering one's life, for example, one's health (life expectancy), financial well-being, personal life, aspirations, career, or long-term security, or the ability to locate and gain assistance toward mastering these tasks. The ability to mobilize defiance, exploit discretion, develop ways of coping, organize protection, and access different knowledge bases is a significant part of such strategies, and therefore of the conviction that one is in charge and not merely the victim of fortuitous circumstances.

The growing knowledgeability of actors in modern societies, or the enhanced bundle of competencies, represents the foundation of the ability of self-organization of small groups of actor in different social roles, for instance, as employees, consumers, tourists, workers, students, or politically active citizens.

CONCLUSION

Our specification of knowledgeability as one of the basic foundations of innovativeness should not be misunderstood to represent causal references to feature of subjects and their social networks that guarantee innovations and therefore can be used in practice as a kind of calculus that if implemented assures novelties in social action. Innovation is not a problem that awaits a solution. Innovation, in the end, is a wicked issue that defies easy solutions.

The pre-eminent feature of social life is its routine, habitual conduct. Lest we despair about this feature of social life, which on first sight inhibits invention/innovation, we should not forget that habitual conduct not only provides the stability of social conduct that is lacking by virtue of the anthropological constant of world-openness of human existence but offers the pre-conditions for innovation. We have to construct and constantly re-construct social institutions that offer the lacking stability, persistence, resilience, and predictability for social conduct. It is within such stable social environments that innovation becomes possible, in the first instance.

NOTES

1. In this contribution we are partly drawing on arguments from a paper in *Mind & Society* (forthcoming).
2. Karl Marx is one of the best-known proponents of a single, deterministic logic of economic and social development. Marx ([1867] 1967, *Capital,* 1, pref. 13) saw uneven economic development in his times, in particular the early emergence and strength of industrial society in Britain, as subject to a single historical law: "the country that is more developed industrially only shows to the less developed, the image of its own future." But the actual development of industrial society in different countries in terms of timing, specificity, composition, and direction of change did not conform to the logic of singular economic development. One cannot observe any "uniformity of sequence, no single way, no law of development. Each of the would-be industrializers, the so-called follower countries, however much influenced by the

Culture and Cognition 35

British experience—to some extend inspired, to some extent frightened or appalled—developed its own path to modernity (Landes, 1998, 236).
3. These observations do not preclude, however, that the supposedly typical rational attitude toward economic conduct found in liberal-capitalist economies is somehow fixed and may not change and develop. It is possible that we are witnessing, at the present time, an increasing "moralization of the market." Economic processes and products, that is, are more and more judged based not on purely rational premises but with reference to ethical convictions. Such a moralization of the market with respect to the products of biotechnology, for example, would represent such a shift and development in the attitudes typical of the capitalist "spirit" (cf. Stehr, 2000; 2008).
4. Talcott Parsons ([1928] 1991,6), in a review of Werner Sombart's *Der moderne Kapitalismus*, expresses his considerable admiration for Sombart's ability to interpret, based on empirical evidence, "a whole epoch of history in such an illuminating and convincing way in terms of one great leading idea. It gives a unity to his presentation which marks a great advance over the entirely disconnected studies of historical facts presented by the historical school proper."
5. As Niklas Luhmann (1982, 221–222) maintains, economic sociology "can only be developed if its approach is overhauled and if it sets out not from a concept of economic society but from a concept of the economy as a subsystem of society." At the same time, this does not preclude the possibility that one assigns "functional primacy" to the economic system since such primacy appears to fall to the societal subsystem that, as Luhmann (1982, 224) also suggests, "can be structured and differentiated from the rest of society with a higher complexity of its own."
6. For an overview of the results of innovation studies cf. Faulkner, 1994, 434–442.
7. One of the first empirical studies of the interdependence of technical innovation and organizational processes and development is Burns and Stalker's *The Management of Innovation* (1961).
8. For Joseph Schumpeter, innovations become a central, if not the main component of the dynamics of economic action. For example, innovations are seen to be more important than is price competition among firms. According to Schumpeter ([1942] 1962, 132), pioneering entrepreneurs who "reform or revolutionize the pattern of production by exploiting an invention or, more generally, an untried technological possibility for producing a new commodity or producing an old one in a new way, by opening up a new source of supply of materials or a new outlet for products, by reorganizing an industry," are at the center of the dynamics of the capitalist system. In Schumpeter's usage, innovations refer to the initial introduction of a new product (thus, product innovation) or system and process (hence, process innovation) into the economy. Although Schumpeter's terminology extends to organisational and managerial innovations, most of the subsequent analyses carried out in economics that pertain to innovations have concentrated on technical innovations or innovations that relate to artefacts. Since Schumpeter makes a sharp distinction between invention and innovation, it becomes evident that his notion of innovation refers not merely to the fabrication of additional knowledge but to incremental knowledge that has been translated into practice (hence practical knowledge) and results in a new product or process. An invention as additional knowledge (or conceptual invention) is knowledge as defined here, namely, a capacity for action.
9. For a recent cultural artefact that blends many of these frameworks into one version of a contemporary e-innovator, see director David Fincher's

The Social Network (2010), a movie that portrays the steps and challenges that Facebook's founder Mark Zuckerberg faced bringing the website to the center of Western online culture. In her movie review for *The New York Times*, Manohla Dargis suggests that the movie portrays the innovator as the "smartest guy in the room," as a person "who remains a strategic cipher," as someone "charmless and awkward in groups larger than one," and as motivated by an emotional storm fueled by a frustrated libido (Dargis, *The New York Times*, September 23, 2010). See also Susan Cain's op-ed titled "The Rise of the New Groupthink," which reproduces these cultural frameworks that structure understandings of innovation and invention. Pitting solitude against collaboration, Cain asserts that research "strongly suggests that people are more creative when they enjoy privacy and freedom from interruption" (Cain, *The New York Times*, January 13, 2012). Indeed, the boundaries distinguishing popular understandings of innovative thinkers from those identified in research blur. Cognitive psychologists suggest that innovative thinkers are more introverted on average, and prefer engaging with concepts to interacting in routine, interpersonal social contexts (see Church and Waclawski 1998).

10. For as Emile Durkheim ([1912] 1965, 479) perceptively observed: »Life cannot wait« (cf. also Gehlen, [1950] 1988, 296–297). In their discussion of expertise and how expertise may be justified, Harry Collins and Robert Evans (2002, 241) advance similar observations about the essential difficulties encountered in the public domain if one would have to wait for expert advice: "Decisions of public concern have to be made according to a timetable established within the political sphere, not the scientific or technical sphere; the decisions have to be made before the scientific dust has settled, because the pace of politics is faster than the pace of scientific consensus formation."

11. As Benoît Godin (2008)—in a history of the concept of innovation—points out, ovation is a term that first appeared in law in the 13th century. It meant renewing an obligation by changing a contract for a new debtor. The term was rarely used in the various arts before the 20th century. In fact, as with imitation and invention, the term was pejorative for a while. Until the 18th century, a "novator" was still a suspicious person, one to be mistrusted."

12. A comparative anthropological analysis of knowledge systems that does not proceed from the assumption of an essentialist hierarchy of knowledge systems with scientific knowledge invariably at the apex of such a stratified figuration, but rather aims to explore both continuities and differences among forms of knowledge, can be found in Watson-Verran and Turnbill (1995).

13. Peter Drucker (1993, 184) observes, however, that initial economic advantages gained by the application of (new) knowledge become permanent and irreversible. What this implies, according to Drucker, is that imperfect competition becomes an constitutive element of the economy. It is the case, of course, that the wide dissemination and application of knowledge beyond the boundaries of the organization that initially gained an edge (as the result of being ahead of its competitors) does not literally lose the now more widely "shared" knowledge since this is one of knowledge's more peculiar properties. Knowledge can be disseminated or sold without leaving the context from which is disseminated or sold. The edge that remains is perhaps best described as an advantage that could be minor but may also be quite significant, based on cumulative learning or the fact that one is able to benefit from the "first-mover-advantage." All of this does not preclude a strategy among firms that attempts to share the benefits from incremental knowledge and innovations in an attempt to reduce the economic risk of investing into

the fabrication of knowledge and in an effort to increase the payoff from innovative products and services. Among other reasons, the difficulties that may be associated with efforts to appropriate benefits from research efforts in private firms is often employed as a standard justification for the public support of science (see Nelson, 1959; Rosenberg, 1990; Pavitt, 1991, 111); or it is argued that the societal returns from basic research efforts are significant and higher than the private returns, justifying public support for such research (Rosenberg, 1990, 165).

14. Karl Mannheim ([1929] 1936) defines, in much the same sense, the range of social conduct generally, and therefore contexts in which knowledge resources play a role, as restricted to spheres of social life that have not been routinised and regulated completely. For, as he observes, "conduct, in the sense in which I use it, does not begin until we reach the area where rationalization has not yet penetrated, and where we are forced to make decisions in situations which have as yet not been subjected to regulation" (Mannheim, [1929] 1936, 102). Concretely to the point and as Mannheim ([1929] 1936, 102) notes: The action of a petty official who disposes of a file of documents in the prescribed manner or of a judge who finds that a case falls under the provisions of a certain paragraph in the law and disposes of it accordingly, or finally of a factory worker who produces a screw by following the prescribed technique, would not fall under our definition of 'conduct.' Nor for that matter would the action of a technician who, in achieving a given end, combined certain general laws of nature. All these modes of behavior would be considered as merely 'reproductive' because they are executed in a rational framework, according to a definite prescription entailing no personal decision whatsoever."

15. Democratized" innovation processes refer to the capacity of users of products and services—both firms and individuals—to innovate for themselves rather than rely on manufacturing-centric innovation (Von Hippel, 2006, 1).

16. The "faculty to decide" already occurs in Montesqueu's treatise on the Spirit of the Laws (Book 11, chapter 6) and reads as follows: "The faculty to decide is what I call the right to issue orders in one's own names or to correct orders issued in someone else's name" (Rosanvallon, [2006] 2008, 121). An alternative translation refers to this faculty as the "power of enacting."

17. The faculty to organize protection resonates with Montesquieu's (political) concept of the "faculty to prevent" (also translated as the "power of refusing"): "The faculty to prevent is what I call the right to nullify a resolution taken by someone else" (this translation from *De l'Esprit des lois* by Arthur Goldhammer can be found in Rosanvallon ([2006] 2008, 121). Although the faculty to organize protection or the faculty to refuse and prevent are, in some sense, "negative" or conserving capacities, the capacity to organize protection calls attention to a resource that may facilitate—in case of inventiveness and innovativeness—premature closure on the process of cognitive displacement as it may originate from conditions of the social context.

18. Sprague and Rudd (1988) have examined the nature and the extent of organizational dissent in high-technology industry.

19. In the United States, the social location of many of these activities can be found in what Peter Drucker (1989,187) calls the ˝third˝ sector of non-profit, non-governmental, ˝human change˝ institutions (or, the "civil society sector" as Salamon and Anheier [1997] have called it). The third sector is actually the "country's largest employer, though neither its workforce nor the output it produces show up in the statistics. One out of every two adult Americans—a total of 90 million people—are estimated to work as volunteers in the third sector" (Drucker, 1989, 197).

REFERENCES

Austin, J.L. ([1962] 1975) *How to Do Things with Words*. Cambridge: Harvard University Press.

Beckenbach, F., & Daskalakis, M. (2010) Invention und Innovation als creative Problemlösungsprozesse: Ein Beitrag zur Mikoröknomik des Wissens. Pp. 259–292, in Manfred Moldaschl and Nico Stehr (eds.), Wissensökonomie und Innovation. Beiträge zur Ökonomie der Wissensgesellschaft. Marburg an der Lahn: Metropolis.

Bourdieu, P. (1975) The specificity of the scientific field and the social vonditions of the progress of reason. *Social Scientific Information*, 14 (6), 19–47.

Broman, T. H. (2002) Some preliminary considerations on science and civil society. *Osiris* 17, 1–21.

Burns, T., & Stalker, G.M. (1961) *The management of innovation*. London: Tavistock.

Church, A. H., & Waclawski, J. (1998) The relationship between individual personality orientation and executive leadership behavior." *Journal of Occupational and Organizational Psychology*, 71 (2), 99–125.

Collins, H. M., & Evans, R. (2002) The third wave of science studies: Studies of expertise and experience. *Social Studies of Science*, 32, 235–296.

Dosi, G. (1984) *Technical change and industrial transformation: The theory and an application to the semiconductor Industry*. London: Macmillan.

Drucker, P.F. (1989) *The new realities: In government and politics/in economics and vusiness/in society and world view*. New York: Harper & Row.

Drucker, P. F. (1993) *Post-capitalist society*. New York: HarperBusiness.

Durkheim, E. ([1912] 1965) *The elementary forms of religious life*. New York: Free Press.

European Commission. (2010) *Innovation. Creating knowledge and jobs*. (Insights from European Research in Socio-Economic Sciences). Brussels: Directorate-General for Research Socio-Economic Sciences and Humanities

Faulkner, W. (1994) Conceptualizing knowledge used in innovation: A second look at the science-technology distinction and industrial innovation. *Science, Technology & Human Values*, 19, 425–458.

Gehlen, A. ([1950] 1988) *Man: His nature and place in the world*. New York: Columbia University Press.

Geroski, P. (1995) "Markets for technology: Knowledge, innovation and appropriability. Pp. 90–131 in Paul Stoneman (ed.), *Handbook of the economics of innovation and technological change*. Oxford: Blackwell.

Gibbons, M., & Johnston, R. (1974) The roles of science in technological innovation. *Research Policy*, 3, 220–242.

Gibbons, M., et al. (1994) *The new production of knowledge: The dynamics of science and research in contemporary societies*. London: Sage.

Godin, B. (2008) Innovation: A history of a category. (Project on the Intellectual History of Innovation, Working Paper No. 1). *http://www.csiic.ca/innovation.html*. Accessed October 5, 2010.

Godin, B. (2010) Innovation without the word: William F. Ogburn's contribution to the study of technological innovation. *Minerva*, 48, 277–307.

John, R.R. (1998) The politics of innovation. *Daedalus*, 127, 187–214.

Landes, D.S. (1998) *The wealth and poverty of nations*. New York: W.W. Norton.

Larson, M.S. (1990) In the matter of experts and professionals, or how impossible it is to leave nothing unsaid. Pp. 24–50 in Rolf Torstendahl and Michael Burrage (eds.), *The formation of Professions. Knowledge, state and strategy*. London: Sage.

Latour, B. (1993) *We have never been modern*. Cambridge: Harvard University Press.

Luhmann, N. (1982) The differentiation of society. New York: Columbia University Press.
Lyotard, J. ([1979] 1984) The postmodern condition: A report on knowledge. Minnesota: University of Minnesota Press.
Marx, K. ([1867] 1967) *Capital: A critique of political economy.* New York: International Publishers.
Mannheim, K. ([1929] 1936) *Ideology and utopia. An introduction to the sociology of knowledge.* London: Routledge and Kegan Paul.
Moldaschl, M. (2010) *Innovation in sozialwissenschaftlichen Theorien oder Gibt es überhaupt Innovationstheorien?* Manuscript.
Nelson, R.R. (1959) The simple economics of basic scientific research. *Journal of Political Economy* 67, 297–306.
Parsons, T. ([1928] 1991) "Capitalism" in recent German literature: Sombart and Weber. Pp. 3–37 in Talcott Parsons, *The Early Essays.* Chicago: University of Chicago Press.
Pavitt, K. (1991) What makes basic research economically useful" *Research Policy,* 20, 109–119.
Rosanvallon, P. ([2006] 2008) *Counter-democracy. Politics in an age of distrust.* Cambridge: Cambridge University Press.
Rosenberg, N. (1990) Why do firms do basic research? *Research Policy,* 19, 165–174.
Salamon, L.M., & Anheier, H.K. (1997) The civil society sector. *Society,* 34 (2), January–February, 60–65.
Saussure, F. ([1916] 1983) Course in general linguistics, (eds.) Bally and Sechehaye. Chicago: Open Court.
Sartori, G. (1968) Democracy. In Davis Sills (ed.), *International encyclopedia of the social sciences,* Volume 4. New York: Macmillan and Free Press, pp. 112–121.
Schieman, S., & Plickert, G. (2008) How knowledge is power: Education and the sense of control. *Social Forces,* 87, 153–183.
Schon, D.A. ([1963] 1967) *Invention and the evolution of ideas.* London: Tavistock.
Schumpeter, J.A. ([1942] 1962). *Capitalism, socialism and democracy.* New York: Harper.
Schumpeter, J.A. ([1911] 1934) The theory of economic development: An inquiry into profits, capital, credit, interest and the business cycle. Cambridge, Mass.: Harvard University Press.
Sombart, W. ([1916] 1921) *Der moderne Kapitalismus. Historisch-systematische Darstellung des gesamten Wirtschaftslebens von seinen Anfaengen bis zur Gegenwart.* Erster Band: Einleitung—Die vorkapitalistische Wirtschaft—Die historischen Grundlagen des modernen Kapitalismus. Erster Halbband. Munich, Leipzig: Duncker & Humblot.
Sprague, J., & Rudd, G. (1988) Boat-rocking in the high-technology culture. *American Behavioral Scientist,* 32, 169–193.
Stehr, N. (2000) *The fragility of modern societies.* London: Sage.
Stehr, N. (2008) *Moral markets.* Boulder, CO: Paradigm Press.
Swidler, A. (1986) Culture in action: Symbols and strategies. *American Sociological Review,* 51 (2), 273–286.
Von Hippel, E. (2006) *Democratizing innovation.* Cambridge: MIT Press.
Vromen, J.J. (2001) The human agent in evolutionary economics. Pp. 184–208 in N.J. Laurent and J. Nightingale (eds.), *Darwinism and evolutionary economics.* Cheltenham: Edward Elgar Publishers.
Watson-Verran, H., & Turnbull, D. (1995) Science and other indigenous knowledge systems. In S. Jasanoff, G. Markle, T. Pinch, & D.J. Petersen (eds.), *Handbook of science and technology studies.* Thousand Oaks, CA: Sage Publications, pp. 115–139.
Woolgar, S. (1998) A new theory of innovation. *Prometheus,* 16, 441–452.

2 Socially Responsible Science and Innovation in Converging Technologies

Toni Pustovrh

INTRODUCTION

Innovations are increasingly regarded as one of the main drivers of national economic growth and competitiveness, with "innovation" having replaced the previously treasured term of "progress" Although a myriad of diverse material and sociocultural elements is crucial for a framework capable of fostering and promoting innovation, scientific and technological (S&T) developments remain the primary source of innovative applications, which can be either social or technological in nature. Many new and emerging fields of S&T are, at least according to some experts and advocates (Kurzweil, 2005; Williams and Frankel, 2007; Roco et al., 2011), increasingly expected to serve as the future source of innovative applications that will (re)fuel the growth of national economies in the coming decades.

Such trends are evident in the European efforts to establish an "Innovation Union" that would, through the comprehensive fostering and use of innovations, ensure the future competitiveness and growth of the European Union (EC, 2012), as well as in the projections of innovation experts who, for example, estimate that more than one-third of the U.S. GDP in 2005 can be attributed to innovation, especially technological innovation, marking such economies as "Innovation Economies" (Canton, 2005, 37). Extrapolations of such trends suggest that the importance of technological innovations in future GDP levels can only increase.

Conversely, the growing power and influence of new and emerging technologies can also carry great risks, ranging from the development of undesirable social trends and disruptive societal transformations to various negative impacts on the environment, health, and safety (Nordman, 2004; Coenen et al., 2009; Schmidt et al., 2009). Such risks have led to declining public trust in S&T, a loosening of social cohesion, and even to "citizen revolts" (Ananda, 2010) against some applications of modern biotechnologies and nanotechnologies.

Through ever more rapid S&T developments and accompanying social transformations, modern societies are experiencing a fragmentation of interests and values and the dissolution of traditional social institutions and networks, along with declining levels of trust in the decisions of experts. In addition, they are facing growing fragility resulting from increasing societal complexity, heterogeneity, and reliance on centralised and risky technologies (Pustovrh, 2010). In such conditions, social capital can be seen as a crucial resource that needs to be strengthened in order to develop and extend the networks of trust and reciprocity so as to facilitate co-operation and mutually supportive relations (Putnam, 1995) among the different groups, actors, and stakeholders involved in the S&T production process in order to increase general societal co-operation and cohesion, as well as to inform scientists, policymakers, and the industry of socially broadly desirable and acceptable technological innovations, thereby making modern societies more resilient and robust.

Such goals have prompted experts and policymakers to seek novel ways of addressing the potential negative implications and risks of new and emerging technologies in the early phase of research and development, and extending the scope of deliberations on S&T developments to a wider range of diverse stakeholders as well as the various publics, while also striving to include specific societal needs and demands in the innovation process (Nordman, 2004; Bainbridge and Roco, 2005; Roco and Bainbridge, 2010; Von Schomberg, 2012). These mechanisms aim to produce what might be termed socially responsible science and innovation; social capital, defined as positive social relations between members of different societal segments that increase co-operation and confidence to obtain collective or economic results, represents a crucial element in fostering the production of "socially robust" science and innovation directed at promoting the "common good" of the entire society.

The first section of this chapter thus provides a brief overview of the new and emerging technologies expected to produce the innovations that could serve as economic and market drivers for the growth and competitiveness desired in developed countries. It also highlights certain risks inherent in their development and points to some socioeconomic factors, such as patent trends, that could shape their role as future innovations.

The second section examines several social governance mechanisms in the framework of socially responsible science and innovation, which could serve to increase public trust and societal resilience regarding the potential risks of new and emerging technologies, while addressing the need to steer technological developments toward pressing problems and needs facing society. It focuses on the role of social capital in increasing cohesion and co-operation within the scientific community or the "society of science" (Mitcham and Stilgoe, 2009), and to promote socially robust science and innovation by engaging and empowering civil society through the facilitating role of civil society organisations (CSOs) as the key mechanisms for communicating the needs, values, concerns, and interests of the citizens or wider society to the society of science, industry, policymakers, and other actors in the S&T production process, resulting in socially beneficial and desired innovations.

Finally, taking into account the innovation potential and dangers of new and emerging technologies as well as the features of the proposed mechanisms for establishing socially responsible science and innovation, this chapter affirms the crucial need to strengthen social capital in the various societal segments involved and influenced by the S&T production process as an important element of modern societal cohesion and stability that has been somewhat neglected in other proposals for the governance of S&T.

CONVERGING TECHNOLOGIES AS RISK SOURCES AND INNOVATION DRIVERS

In the early years of the 21st century, a group of scientists, experts, business leaders, and National Science Foundation officials from the United States formulated a new paradigm for technological innovation in the new millennium, namely, the concept of Converging Technologies (CTs) defined as innovative technologies emerging from the synergistic combinations and mutual stimulation of scientific and technological developments among the four expansive domains of Nanotechnology, Biotechnology, Information Technology, and Cognitive Science[1] (NBIC) (Roco and Bainbridge, 2003, ix). A subsequent EU report on Converging Technologies for the European Knowledge Society defined CTs as breakthrough technologies and knowledge systems that enable each other in the pursuit of a common goal (Nordmann, 2004, 14). The new and emerging technologies that are being formed at the intersections of the NBIC Converging Technologies are expected by many experts to be the source of major economic innovations that will drive the growth of knowledge societies (Kurzweil, 2005; Williams and Frankel, 2007; Allhoff et al., 2010; Roco et al., 2011) and produce applications that will profoundly expand the current spectrum of human performance, solve many physical and societal problems, and transform individuals, groups, and societies for the better.

Other experts (Nordmann, 2004) have pointed out that the development and commercialisation of CTs could also pose great risks to the environment, human health, and safety and endanger many current belief systems and values and the stability of numerous social structures and institutions, as well as entire societies. Some have also warned that technologically increased or enhanced individual and societal capabilities would not necessarily contribute to increased happiness (Glazer, 2006; Kass, 2008), living a "good life" (Sandel, 2007), or solving pressing problems such as inequality, at least not without suitable accompanying social policies (Hughes, 2004), as they might produce undesirable social trends and disruptive societal transformations in terms of societal access, fairness, and justice (Allhoff et al., 2010). The weaponised use of some new and emerging technologies could even result in global catastrophic risks, events capable of causing extensive loss of life and/or massive economic damage (Bostrom and Ćirković, 2008).

Socially Responsible Science and Innovation   43

Despite some of the alarming potential risks of CT applications, their development is still likely to continue given the importance of S&T developments in the modern innovation-oriented and market-driven economies that often follow the proactionary developmental model characterised by free entrepreneurship, technology-driven innovation, and minimal regulation characteristic of the United States.[2]

Since patents are often regarded as one of the indicators of S&T fields that are expected to produce commercially attractive innovations, the continuation of this section examines patent trends[3] in the period between 1999 and 2007 for the CT domains of nanotechnology, biotechnology, and information-communication technology.[4]

As the summary of the patent trend analysis in Table 2.1 shows, the number of patent applications and grants in the first three CT (nanotechnology, biotechnology. and ICT) domains has steadily declined from a peak number reached during the examined period.

The number of worldwide patents in nanotechnology in 2007 at the EPO was 65% (896) of the peak (1,376) in 2004, and 15% (359) of the peak (2,416) in 2002 at the USPTO.

The number of worldwide patents in biotechnology in 2007 at the EPO was 75% (6,845) of the peak (9,093) in 2000, and 11% (912) of the double peak (8,398 and 8,359) in 1999 and 2000 at the USPTO.

The number of worldwide patents in ICT in 2007 at the EPO was 89% (40,622) of the peak (45,428) in 2004, and 21% (21,140) of the peak (99,236) in 2000 at the USPTO.

Judging from the fluctuation of patent numbers alone, one might conclude that the innovative potential of the first three CT domains has reached its high point and is now in decline. As there is a finite number of innovations that are worth patenting in a given segment and period of scientific and technological development, one explanation might be that the "patenting potential" of individual Nanotechnology, Biotechnology, and ICT domains has been declining since the individual peaks reached in the

Table 2.1 Worldwide Patenting Trends in Nanotechnology, Biotechnology, and ICT at the EPO and USPTO

Year			1999	Peak Year		2007	
Patents		%	No.	100%	No.	%	No.
Nanotechnology	EPO	54	746	2004	1,376	65	896
	USPTO	71	1,715	2002	2,416	15	359
Biotechnology	EPO	93	8,420	2000	9,093	75	6,845
	USPTO	100	8,398	1999	8,398	11	912
ICT	EOP	87	39,545	2004	45,428	89	40,622
	USPTO	92	90,954	2000	99,236	21	21,140

examined period, at least up until 2007, and new discoveries and convergences are needed to spur a new rise in the number of potential innovative applications and new patents. Additional contributing factors of such patent trends can also be sought among specific conditions in the relevant innovation systems and the wider socioeconomic frameworks, especially in groundbreaking research projects and public funding initiatives, potential benefits and risks, regulatory positions toward and societal acceptance of technological innovations, which in turn again stimulate or impede scientific and technological progress.[5]

An exception to the declining patenting trend is the fourth CT domain of neurotechnology where the number of neurotechnology-related patents is actually increasing, although the numbers only pertain to patent claims and grants at the USPTO. The number of neurotechnology patents granted, identified as drugs, devices, and diagnostics for the brain and nervous system, was increasing steadily over the period from 1985 to 2009, with a sharp rise especially in the past few years. Neurotechnology patent claims filed with the USPTO have increased 200% compared with the 60% growth in total patent claims over the past 10 years (Lynch et al., 2010, 7). Currently, the value of the existing and emerging neurotechnology industry is estimated at U.S.$ 143 billion, with more than 800 companies being involved in the manufacture and marketing and/or research and development of various neurotechnologies (NeuroInsights, 2010).

Thus, judging from the patent trends of the four domains of CT the area of neurotechnology seems to be the most promising source of new and emerging technologies that could drive the economies of developed countries in the future, yet it is also a domain of applications that could be ethically and societal contentious because of their inherent risks and disruptive social impacts.

SOCIALLY RESPONSIBLE SCIENCE AND INNOVATION AND SOCIAL CAPITAL

Throughout modern history, S&T have played an increasingly important role as catalysts of cultural and social change, affecting and shaping human societies and individuals. As many highly developed industrialised countries claim to be knowledge societies (Stehr, 2001) in which knowledge is the primary production resource, created, shared, and used for the prosperity and well-being of the citizens, scientific exploration of the workings of nature and the development of technologies employing such knowledge continue to occupy a central role in maintaining competitiveness and economic growth.

On the other hand, scientific insights and technological innovations have also resulted in an accelerated dismantlement and transformation of traditional institutions and social relationships that have long provided stability, co-operation, and cohesion in preindustrial societies (Beck, 1992). The changing structure and function as well as increasing complexity of modern societal mechanisms have resulted in the growing role occupied by scientific and technological developments, accompanied by the rise of diverse expert bodies tasked with the role of managing individual societal subsystems. While specialised experts are indispensable in the working of modern technological societies, the overreliance on the objectivity and unbiased knowledge of experts has resulted in the exclusion of "laymen" citizens and a disregard for their needs, concerns, values, and interests in the S&T production process. Coupled with concerns regarding the possible intended (hostile) or unintended negative consequences of new and emerging technologies, this has led to a general drop in public trust in experts, as well as in the proclaimed "best public interest" of decisions made by the society of science, industry, and policymakers. Such developments have recently even resulted in "citizen revolts" against the introduction of technological innovations, such as genetically modified food crops, and nanoparticles and nanomaterials, which were developed and deployed without any wider public input or deliberation and even in the face of widespread public opposition (Ananda, 2010).

As modern technological societies are becoming increasingly complex and diversified, the role of social capital, especially as manifested in consensus-building leading to collective action and implying a shared interest and agreement in society (Arefi, 2003), is becoming a key element of fostering social cohesion and resilience in the light of the societal fragility resulting from a growing reliance on complex technology and highly specialised expertise. The pluralistic fragmentation of values and common interests in modern societies, strengthened with declining trust in the objectivity and reliability of expert knowledge and concerns about the growing negative health, environmental, and socially disruptive effects of new and emerging technologies, has prompted calls for extensive bans and even the complete relinquishment of entire domains of S&T development, such as genetics, robotics, nanotechnology, and machine intelligence (Joy, 2000; McKibben, 2004; Kass, 2008).

A reduction of the complexity and plurality of modern societies seems quite unrealistic and even undesirable given the creative and innovative potential entailed in diversity and freedom, which is confirmed by the view of social capital as promoting both the importance of community in building generalised trust and in the importance of individual free choice in creating a cohesive society (Ferragina, 2010). Furthermore, the new and emerging technologies that have been identified by technology experts (Mulhall, 2002; Garreau, 2005; Kurzweil, 2005) as the drivers of future economic and societal development overlap with the domains mentioned by the opponents to further development of the S&T mentioned above.

Accordingly, novel approaches have turned to the development and implementation of various societal mechanisms that would enable socially responsible and responsive science and innovation, addressing the need for the early detection and mitigation of potential negative consequences, as well as the need to broaden the scope of deliberation on the nature and impact of S&T products, including a broader range of stakeholders and the wider society.

Several of these mechanisms also aim to strengthen social capital, either within the society of science itself or in the connection between the society of science, wider society, and the other actors in the S&T production process. Several scientific communities working in areas of the new and emerging technologies have attempted to ensure safety and responsibility through self-regulation by establishing and adopting internal codes of ethics and conduct. For example, guidelines for increasing safety and security in the field of synthetic biology through community self-regulation have been proposed (Maurer et al., 2006), with goals similar to those of the Asilomar conference in 1975 for genetic engineering, which yielded a temporary moratorium on some experiments until their safety could be further assessed. While the international Synthetic Biology 2.0 meeting of the main actors in the domain rejected the proposed guidelines, arguing that the field should be subject to further public debate and government regulation and that it was too early to impose the limiting protocols set out in the guidelines (Aldhous, 2006), they did agree to develop software protocols for the detection of dangerous genetic sequences and to report suspicious experiments, fostering internal codes of ethics and conduct based on self-governance. Sophisticated codes of ethics and conduct were also rapidly adopted among the growing amateur Do-It-Yourself Biology movement, a loose global community of amateur citizen scientists, which emerged in recent years because of the falling costs and increasing availability of molecular biology equipment (DIYBio, 2012). As a report on the global governance of science (Mitcham and Stilgoe, 2009) suggests, the globalisation of standards and shared codes of ethics and conduct within the scientific and research communities worldwide would be the best protection against unethical research and development, which might otherwise be conducted in countries where regulation is non-existent or easily gives way to potential profits, in so-called ethics-free zones.

Such adoptions of shared codes of ethics and conduct within different scientific communities could strengthen the social capital within the society of science by increasing commitment to shared values and ethics, promoting internal trust and reciprocity. While this is an important step in ensuring the safety and security of the research and its results, and in enabling greater interdisciplinary co-operation between diverse communities of scientists, strengthening bridging social capital between the society of science and other segments of society will be necessary to prevent the society of science from remaining self-contained and isolated, with its own values, norms, and interests which might not necessarily overlap with the needs, values, and interests of the other stakeholders and wider society. The problem of scientists becoming increasingly remote from the common interests and needs they are supposed to address has been recognised at least since the beginning of the 20th century (Dewey, 1988), and can be seen as a negative manifestation of bonding social capital. The society of science is already to some extent performing bridging activities through individual scientists and scientific organisations that are concerned with fostering the public engagement of science in society, educating and informing wider society on the scientific method, novel insights, research and development, as well as socially oriented endeavours, thereby increasing bridging social capital between the society of science and wider society.

The framework of socially responsible science and innovation further entails the promotion of the "right impacts" of S&T development to steer research and development through two important elements recently summarised under the concept of Responsible Research and Innovation (Sutcliffe, 2011; Von Schomberg, 2012). Concerned about the increasing doubts amongst the general public in developed countries during the last decade regarding the objectivity and reliability of scientific expertise, particularly on the societal benefits, risks, and unintended consequences of new technologies, two further mechanisms that might enable greater responsibility and responsiveness of science have been proposed. The first is the "upstream" investigation of the potential Ethical, Legal and Societal Implications (ELSI) of new emerging technologies, and the second is enabling the participation of a wide range of stakeholders, including the general public, in deliberations regarding technological development, which also encompasses the need for the greater social responsiveness of S&T.

The investigation of the potential ramifications of technological development while the technology is still in early research and development stages first became widely visible as part of the Human Genome Project that investigated the possible ethical and societal implications of sequencing the entire human genome. Since then, ELSI investigations have been conducted in several fields of the new and emerging technologies (Nordmann, 2004; Bainbridge and Roco, 2005), for example synthetic biology (Schmidt et al., 2009) and neurotechnology (Allhoff et al., 2010; Coenen et al., 2009), with the aim of highlighting potential negative impacts of such innovations on the individual and society, and attempting to address and mitigate them even before technological deployment.

While the impact assessments done by social scientists and philosophers have largely focused on the potential negative consequences of the new technologies, and the work of the natural scientists focused mostly on promoting their benefits, there is a need for co-operation between the "two cultures" of the natural sciences and the arts and humanities (Snow, 1998) in joint projects, where both sides work together with the goal of finding ways to prevent or mitigate the negative and realise the positive impacts of novel S&T developments. One such example is the UK Centre for Synthetic Biology at Imperial College London where social scientists have been integrated into the research team (Sanderson, 2009). Co-operation and joint projects could thus further strengthen the social capital and cohesion and co-operation in the society of science, especially between the two cultures that are almost traditionally antagonistic.

Just as there is a need for dialogue and greater co-operation between the natural sciences and the arts and humanities, there is also an ever stronger need for dialogue between general society, that is, civil society and other stakeholders on one side and the society of science, industry, policymakers, and other actors in the S&T production process on the other. The social capital between these societal segments could be strengthened by the development and fostering of tentative mechanisms for stakeholder and public participation in science governance that are oriented toward a democratic turn in active citizen participation in S&T (Mejlgaard and Stares, 2010), characterised by openness, transparency, and deliberation. Ideally, such mechanisms would enable various stakeholders as well as the general public to express their opinions, share their expertise and knowledge, and provide a two-way channel between science and society that would in turn also strengthen public trust by incorporating and addressing the needs, concerns, and expectations of general society into novel technologies. A further aspect of enabling participation would be ensuring the greater social responsiveness of science not only to general societal needs, but also to national, regional, and local needs, taking into account the local culture, problems, concerns, and specific risks (Jimenez, 2008), as opposed to citizens and societies being assigned the role of passive consumers and recipients of externally developed and imposed innovations.

By developing such channels of communication and networks of relationships, the bridging social capital between the various mentioned societal segments and consequently society as a whole would be strengthened, manifested in increased civic and stakeholder action and communication of needs, concerns, and interests, as the citizen would feel that the common and particular interests and values are actually being taken into account and are being adequately addressed in the S&T production process.

All of these novel requirements regarding knowledge production systems in modern societies point to a strong need to develop "hybrid forums" (Callon et al., 2009) where scientists, experts, policymakers, and citizens discuss and create new approaches to the social regulation of S&T. While there has been much talk on the need to develop such deliberative mechanisms and forums, a recent overview of National Ethics Committees (NECs) in 32 European countries, expert bodies charged with the mission of providing expert opinions on the ethics and other impacts of novel, especially medical, procedures and technologies, which are ideally situated to integrate participatory mechanisms between science and society, shows that only a minority of these institutions have developed active, two-way participatory mechanisms (Mali et al., 2011).) Fewer than half of the 32 NECs (15) feature some type of public involvement mechanisms. Furthermore, a majority of these (11) feature only passive mechanisms, one-way channels of information flow from the experts to the public with the purpose of informing and educating, and only a minority (4) feature active mechanisms, namely, two-way channels that enable the exchange of knowledge, preferences, and opinions between experts on one side and civil society and various stakeholders on the other. Only the National Ethics Committees in Germany, the Netherlands, Portugal, and the United Kingdom have such active participatory mechanisms, with solely the United Kingdom's Nuffield Council on Bioethics performing extensive participatory consultation with stakeholders and the general public, including such gathered views in their opinion documents.

As the number of expert institutions, such as NECs, that have even begun to develop participatory mechanisms is still extremely low, and a reliance on expert knowledge and advice in the S&T production process still predominates, there is an increasing need and opportunity for CSOs to take up the role of forming bottom-up interface mechanisms between wider society and the society of science, industry, and policymakers, especially as CSOs can be more in touch with the local needs, culture, and concerns. In the scope of the socially responsible science and innovation approaches that include deliberations with a wide range of stakeholders and especially the various segments of wider society as key elements, CSOs could occupy important functions. The Global Governance of Science report (Mitcham and Stilgoe, 2009) proposes several measures whereby CSOs could act as important mediating and connecting elements. Research projects that enact fundamental human rights might be strengthened by greater co-operation with CSOs who promote human and citizen rights. Making the results of research as widely available as possible would profit from collaboration with open source initiatives and CSOs engaged in promoting the wide dissemination of knowledge and tackling intellectual property rights issues. As local culture, values, knowledge, and specific local problems are increasingly recognised as important elements to be taken into account when developing and introducing new technologies, the role of CSOs that are familiar with the conditions and operating at a specific local or interest level becomes even more important, especially when they can provide valuable insights that are not available at the higher expert level.

Although a greater plurality of viewpoints and opinions could undoubtedly result in socially more acceptable and robust innovations, the efforts to include the knowledge of "lay experts" do not mean that the casual opinions and views of every citizen, "the man on the street," are on a par with expert knowledge. Instead, the type of citizen needed for participation in the mechanisms of socially responsible science and innovation is an active citizen, an educated and informed participant broadly cognizant with the knowledge concerning the new and emerging technologies, a "well-informed citizen" whose knowledge is derived from consultations with eyewitnesses, insiders, analysts, and commentators (Schutz 1946), and, in modern times, by making selective and critical use of the plethora of information available online.[6] It thus falls to the CSOs to ensure that their members are able to meaningfully participate and communicate their concerns and needs in the deliberations with the other actors and stakeholders in the S&T production process.

While there is a clear need for some type of societal steering regarding the developmental trajectories of S&T research, especially considering the potential of the new and emerging technologies for causing wide ranging negative impacts and unintended consequences, extensive restrictions and too rigidly goal-directed research imposed by actors outside the society of science could also lead to a reduction of the autonomy of scientific research, which could lead to the stagnation of scientific knowledge production and consequent innovation (Bostrom, 2007). Studies (van Lieshout et al., 2006) show that while the EU is ahead of the United States in the production of CT-related knowledge, it lags behind in commercialising such knowledge.[7] This might be related to the EU's (at least in principle at the supranational level) favouring the precautionary principle approach and the United States's favouring the proactionary principle approach. A less innovative and too cautious approach might thus also result in the loss of key industries, missed developmental opportunities, the loss of employment, economic opportunities, and growth drivers, which could manifest in wider societal disruption, mass unemployment, declining standards of living, and a loss of societal cohesion, finally causing a decrease of social capital.

The best approach would thus be to base the goals of S&T impacts on consensual values, needs, and interests, arrived at through wide societal deliberation fostered by CSOs and other measures that strengthen social capital, and ongoing processes of ELSI examination that balance proaction and precaution and seek ways to mitigate the negative and increase the positive impacts of new and emerging technologies.

CONCLUSION

The impacts and transformative influences of modern S&T developments are too extensive and disruptive not to be the subject of scrutiny and consensus of the whole of society which is directly influenced and transformed by new and emerging technologies. There is thus a strong need to reach a broad societal consensus on the acceptable and desirable developmental trajectories for S&T innovations, based on widely shared values and norms, and strengthened by networks of reciprocity and two-way communication channels between all the actors and stakeholders involved in and affected by the S&T production process.

As we have endeavoured to show, given its innovation potential, neurotechnology could become one of the new and emerging technologies that might serve as a driver of economic development and competitiveness in developed countries, but it also has the potential to be an ethically and socially contentious technological domain. Neurotechnological applications could exacerbate the issues of justice, fairness, and accessibility (Glazer, 2006; Allhoff et al., 2010; Coenen et al., 2009), resulting in negative social changes and trends that primarily affect wider society, that is, the citizens, as well as decrease the resilience and stability of society as a whole. Thus the promotion of mechanisms that strengthen social capital and wide societal cohesion through the formation and extension of networks of trust and reciprocity among the various actors and stakeholders in the S&T production process can be seen as an integral component of any socially responsible approach to science and innovation governance.

Considering that modern technological societies are indeed risk societies where the technological risks, which are systematically produced in the modernisation process and increasingly endanger life on Earth, are an inherent feature of the technological and economic development in the social and production systems of modernity (Beck, 1992), the efforts to develop socially robust innovation processes that balance proaction and precaution and integrate ethical and societal acceptability and desirability, as attained through wide societal deliberation and agreement, are clearly important.

While several examinations (Roco and Bainbridge, 2003; Nordman 2004; Bainbridge and Roco, 2005; Allhoff et al., 2010; Coenen et al., 2009; Schmidt et al., 2009; Roco and Bainbridge, 2010; Roco et al., 2011) of the potential ramifications of the various new and emerging technologies from the domains of CT have been undertaken in the last decade, few if any focus on the importance of social capital both as bonding social capital within the society of science, that is, between various scientific communities and especially between the "two cultures" of the natural sciences and the arts and humanities, and on bridging social capital between wider society, the society of science, policymakers, industry, and other actors and stakeholders in the S&T production process.

The main contribution of this chapter thus lies in highlighting the importance of social capital in the framework of socially responsible science and innovation to foster societal resilience, cohesion, and stability in light of the risks and innovative potential of new and emerging technologies forming in the framework of CT.

The first area focuses on forming and extending the networks of trust and reciprocity between individual scientists and scientific groups within the society of science, through the adoption of codes of ethics and conduct and fostering of commitment to shared values and norms, which could increase the safety and security of the research and its results, and enable greater interdisciplinary co-operation among different communities of scientists.

The second area focuses on strengthening the co-operation and reciprocity between the "two cultures" of the natural sciences and the arts and humanities in order to develop integrated ELSI investigations in the early phase of S&T development through mutual projects in which all scientists involved focus on all salient impacts and strive to form consensual developmental approaches based on common goals and values, striving for impacts that promote the "common good" of society at large.

The third area focuses on creating and expanding the networks of co-operation and exchange between the wider society on one side and the society of science and the other actors in the S&T production process on the other. While there has been much talk about the need to develop hybrid forums (Callon et al., 2009), deliberative platforms where scientists, experts, policymakers, and citizens can discuss and create joint approaches to the governance of S&T, the number of expert institutions such as national ethics councils that have actually developed active mechanisms for stakeholder and public participation is still extremely low (Mali et al., 2011). This situation opens up the opportunity for civil society organisations to empower civil society and stimulate the civic engagement and action of wider society on S&T issues. Such wide societal deliberation and inclusion in the science and innovation process could promote the social responsiveness of science not only to general societal needs, but also to local needs, taking specific local culture and problems into account (Jimenez, 2008). Using CSOs to increase bridging social capital throughout general society and among the various societal segments by communicating societal needs, concerns, and interests could strengthen societal cohesion as the citizen would feel that the common good and particular interests and values are actually being taken into account and adequately addressed by the other actors in the S&T production process. But it should also be noted that employing CSOs in this way requires active and well-informed citizens who are familiar with the various aspects and impacts of the new and emerging technologies as well as the S&T production process.

Since low levels of social capital, seen as the collective value of all social networks and their potential for reciprocity, result in low levels of trust in government and in low civic participation (Putnam, 1995), and the elements in the framework of socially responsible science and innovation could increase social capital through the mechanisms described above, such a framework could represent a "middle way" approach between the relinquishment and uncritical adoption of disruptive technologies as it attempts to strike a balance between innovation and precaution, while integrating the opinions, needs, and concerns of a broad array of stakeholders and wider society, assigning an important role to CSOs as the interface mechanisms between wider society on one side and the society of science, policymakers, industry, and other stakeholders in the S&T production process on the other.

Such mechanisms could promote the common societal "quest for the common good" as suggested in the concept of Responsible Research and Innovation by steering the society of science in co-operation with other stakeholders and wider society to pursue the "right impacts" anchored in the values set out in the Treaty on the European Union, through the pursuit of ethically acceptable, sustainable, and socially desirable elements in both the innovation process and in marketable products (Von Schomberg, 2012). In this framework where both scientists and citizens deliberate on desired S&T trajectories that would serve to increase the common good, all actors and stakeholders involved need to consider both positive and negative impacts and find ways that both might be improved, instead of focusing only on either warning of the risks and negative impacts or solely promoting the benefits and positive impacts of new and emerging technologies. Strong social capital is necessary for democracy and economic growth (Fukuyama, 1999), and shared norms and values among different actors and stakeholders regarding the S&T production process could promote societal co-operation and cohesion, mitigating the modern value and interest fragmentation in increasingly complex and pluralistic societies. Thus fostering and promoting mechanisms of socially responsible science and innovation, which could prevent the negative impacts and costs of the societal rejection of novel technologies imposed without societal deliberation or consent, is crucial in the current time of global crisis and waning economic development and innovation, when new and emerging S&T breakthroughs are needed to promote further socioeconomic development and address pressing societal needs and problems.

NOTES

1. More precisely labeled as neurotechnology, which is closer to the "technological" and "applied" classification nature of the first three domains.
2. Although the EU has, at least on the supranational level, struggled to shape and steer developments in accordance with its goals and values in a precautionary innovation approach (Von Schomberg, 2012).
3. Based on the yearly numbers of patent applications filed with the European Patent Office (EPO) and patent grants at the U.S. Patents and Trade Office (USPTO). The data were extracted from the OECD Statistics Extracts datasets (OECD 2011).
4. Regarding the fourth CT domain, cognitive science, or neurotechnology, no unified category currently exists in the major patent databases as the patents pertaining to various new and emerging neurotechnologies are scattered across several other patent categories. The patent trends in neurotechnology were estimated through the use of neurotechnology industry reports (NeuroInsights, 2010).
5. A final note on patents as enablers and indicators of innovation should also mention critical authors who argue that, at least in some fields of science and technology, patents are being increasingly used to secure and monopolise specific discoveries, to prevent competitors from developing new applications that could rival the patent holders' existing applications, without any intention of actually using the patents to develop new and improved versions. Several patents, especially in the fields of life sciences and biotechnology, have been used as "static protection" and are not being actively pursued (Koepsell, 2009).
6. The modern tools of electronic communication could also be used to develop virtual social capital, which is produced through online social networks.

7. This is to some degree reflected in the larger number of patents at the USPTO as compared to the number of patents at the EPO.

REFERENCES

Aldhous, P. (2006) Synthetic biologists reject controversial guidelines. *New Scientist*. http://www.newscientist.com/article/dn9211-synthetic-biologists-reject-controversial-guidelines.html. Accessed January 30, 2012.

Allhoff, F., Lin, P. Moor, J. & Weckert, J. (2010) *Ethics of Human Enhancement: 25 Questions & Answers*. Studies in Ethics, Law, and Technology, 4 (1), 1–39.

Ananda, R. (2010) GMO crop sabotage on the rise: French citizens destroy trial vineyard. *Food Freedom*. http://foodfreedom.wordpress.com/2010/08/16/gmo-crop-sabotage-on-the-rise-french-citizens-destroy-gmo-vineyard. Accessed January 25, 2012.

Arefi, M. (2003). Revisiting the Los Angeles Neighborhood Initiative (LANI): Lessons for planners. *Journal of Planning Education and Research*, 22 (4), 384–399.

Bainbridge, W.S., & Roco, M.C. (eds.). (2005) *Managing nano-bio-info-cogno innovations: Converging technologies in society*. Dordrecht: Springer.

Beck, U. (1992) *The risk society: Towards a new modernity*. London: Sage Publications.

Bostrom, N. (2007) Technological revolutions: Ethics and policy in the dark. In *Ethics of east and west: How they contribute to the quest for wisdom*, ed. Julian Savulescu. Oxford: Oxford Uehiro Center for Practical Ethics.

Bostrom, N., & Ćirković, M. (eds.). (2008) *Global catastrophic risks*. Oxford: Oxford University Press.

Callon, M., Lascoumes, P. & Barthe, Y. (2009) *Acting in an Uncertain World*. Cambridge and London: The MIT Press.

Canton, J. (2005) NBIC convergent technologies and the innovation economy: Challenges and opportunities for the 21st century. In *Managing nano-bio-info-cogno innovations: Converging technologies in society*, ed. William Sims Bainbridge and Mihail C. Roco, 33–45. Dordrecht: Springer.

Coenen, C., Schuijff, M., Smits, M., Klaassen, P., Hennen, L., Rader, M., & Wolbring, G. (2009) *Human enhancement*. Brussels: European Parliament, DG Internal Policies STOA.

Dewey, J. (1988) *The public and its problems*. Athens: Ohio University Press.

DIYBio. (2012) An institution for the do-it-yourself biologist. http://diybio.org. Accessed January 29, 2012.

European Commission. (2012) *Innovation Union*. http://ec.europa.eu/research/innovation-union/index_en.cfm?pg=keydocs. Accessed August 18, 2011.

Ferragina, E. (2010) Social capital and equality: Tocqueville's legacy. Rethinking social capital in relation with income inequalities. *The Tocqueville Review* 31 (1), 73–98.

Fukuyama, F. (1999) Social capital and civil society. Paper for the IMF conference on Second Generation Reforms. http://www.imf.org/external/pubs/ft/seminar/1999/reforms/fukuyama.htm. Accessed January 23, 2012.

Garreau, J. (2005) *Radical evolution: The promise and peril of enhancing our minds, our bodies—and What it means to be human*. New York: Doubleday & Company.

Glazer, S. (2006) *Enhancement: A cross section of contemporary ethical debate about altering the human body*. Garrison, NY: Hastings Center.

Hughes, J.J. (2004) *Citizen cyborg: Why democratic societies must respond to the redesigned human of the future*. Cambridge, MA: Westview Press.

Jimenez, J. (2008) Research socially responsible: May we speak of a mode 3 knowledge production? *Electronic Journal of Communication Information & Innovation in Health* 2 (1), 48–56.

Joy, B. (2000) Why the future doesn't need us. *Wired*, April 8. http://www.wired.com/wired/archive/8.04/joy_pr.html. Accessed April 23, 2011.

Kass, L.R. (2008) Defending human dignity. In *Human dignity and bioethics: Essays commissioned by the President's Council on Bioethics*. The President's Council on Bioethics, 297–331. Washington, DC: U.S. Independent Agencies and Commissions.

Koepsell, D.R. (2009) *Who owns you: The corporate gold rush to patent your genes*. West Sussex: Wiley-Blackwell.

Kurzweil, R. (2005) *The singularity is near: When humans transcend biology*. New York: Penguin Group Inc.

Lynch, Z., McCann, Corey M., Lynch, Casey Crawford, & Rasmus, Tim. (2010) *Neurotech clusters 2010: Leading regions in the global neurotechnology industry, 2010—2020*. San Francisco: NeuroInsights.

Mali, F., Pustovrh, T., & Groboljšek, B. (2011) *Policy impacts of ethical advisory bodies on the societal regulation of biotechnology*. 10th Annual IAS-STS Conference "Critical Issues in Science and Technology Studies," Graz, May 2–4, 2011.

Maurer, S. M., Lucas, Keith V., & Terrell, S. (2006) *From understanding to action: Community-based options for improving safety and security in synthetic biology*. http://gspp.berkeley.edu/iths/Maurer%20et%20al._April%203.pdf. Accessed January 30, 2012.

McKibben, B. (2004) *Enough: Staying human in an engineered age*. New York: Holt Paperbacks.

Mejlgaard, N., & Stares, S. (2010) Participation and competence as joint components in a cross-national analysis of scientific citizenship. *Public Understanding of Science* 19 (5), 545–561.

Mitcham, C., & and Stilgoe, J. (Rapporteur) (2009) *Global governance of science—Report of the Expert Group on Global Governance of Science to the Science, Economy and Society Directorate, Directorate-General for Research, European Commission*. Luxembourg: Office for Official Publications of the European Communities.

Mulhall, D. (2002) *Our molecular future: How nanotechnology, robotics, genetics and artificial intelligence will transform our world*. New York: Prometheus Books.

NeuroInsights. (2010) *The neurotechnology industry 2010 report*. San Francisco: NeuroInsights.

Nordmann, A. (Rapporteur) (2004) *Converging technologies—Shaping the future of European societies*. Luxembourg: Office for Official Publications of the European Communities.

Organization for Economic Cooperation and Development (2011) *OECD statistics extracts*. http://stats.oecd.org/index.aspx. Accessed January 24, 2011.

Pustovrh, T. (2010) The RISC potential of converging technologies. In *Modern RISC-societies: Towards a new paradigm for societal evolution*, ed. Lučka Kajfež-Bogataj, Karl H. Mueller, Ivan Svetlik, & Niko Toš, 297–324. Vienna: Echoraum.

Putnam, R.D. (1995) Bowling alone: America's declining social capital. *Journal of Democracy* 6 (1), 65–78.

Roco, M.C., & Bainbridge, W.S. (eds.) (2003) *Converging technologies for improving human performance: Nanotechnology, biotechnology, information technology and cognitive science*. Dordrecht: Springer.

Roco, M.C., & and Bainbridge, W.S. (eds.) (2010) *Societal implications of nanoscience and nanotechnology*. Dordrecht, Boston, London: Kluwer Academic Publishers.

Roco, M.C., Mirkin, C.A., & Hersam, M.C. (2011) *Nanotechnology research directions for societal needs in 2020. Retrospective and outlook*. Dordrecht, Heidelberg, London, New York: Springer.

Sandel, M.J. (2007) *The case against perfection: Ethics in the age of genetic engineering*. Cambridge: Belknap Press of Harvard University Press.

Sanderson, K. (2009) Synthetic biology gets ethical. *Nature*. http://www.nature.com/news/2009/090512/full/news.2009.464.html. Accessed January 23, 2012.

Schmidt, M., Kelle, A., Ganguli-Mitra, A., & de Vriend, H. (eds.) (2009) *Synthetic biology: The technoscience and its societal consequences*. New York, Heidelberg: Springer.

Schutz, A. (1946) The well-informed citizen. An essay on the social distribution of knowledge. *Social Research*, 13 (4), 463–478.

Snow, C.P. (1998) *The two cultures*. Cambridge: Cambridge University Press.

Stehr, N. (2001) *The fragility of modern societies. Knowledge and risk in the information age*. London: Sage.

Sutcliffe, H. (2011) *A report on responsible research & innovation*. http://ec.europa.eu/research/science-society/document_library/pdf_06/rri-report-hilary-sutcliffe_en.pdf. Accessed January 30, 2012.

Van Lieshout, M., Enzing, C., Hoffknecht, A., Holtmanspotter, D. (Ed.) Noyons and Ramón Compañó. (2006) *Converging applications for enabling the information society and Prospects of the convergence of ICT with cognitive science, biotechnology, nanotechnology and material sciences*. IPTS report.

Von Schomberg, R. (2012) Prospects for technology assessment in a framework of responsible research and innovation. In Marc Dusseldorp and Richard Beecroft (eds.),. *Technikfolgen abschätzen lehren: Bildungspotenziale transdisziplinärer Methoden*. Wiesbaden: VS Verlag.

Williams, E.A., & and Frankel, M.S. (2007) *Good, better, best: The human quest for enhancement. Summary report of an invitational workshop*. Scientific Freedom, Responsibility and Law Program of the American Association for the Advancement of Science.

3 Culture Impact on Innovation
Econometric Analysis of European Countries

Pedro Ferreira, Elvira Vieira, and Isabel Neira

INTRODUCTION

The globalization of the market economy puts a lot of pressure on companies, regions, and countries, in order to develop new and more aggressive strategies to cope with these demands. Innovation is considered an important path to promote the required competitiveness. However, innovation includes a complex set of processes that involve many actors, and, as a result innovation connects to a wide range of contextual variables.

Many of the processes and techniques derived from innovation efforts are cumulative and interdependent. Much of the ability of a country to develop these efforts stems from the influence of external factors such as the quality of education, especially in infrastructure-related knowledge (universities, research institutes, among others), the availability of infrastructures to support research, and the healthy functioning of markets. These innovation efforts may be reflected on the competitiveness of countries mainly in two ways. The first may result in changes at the level of organisation, methods of production, or marketing strategies to improve the efficiency of the entire production or commercial process; the second can lead to the introduction of new products and the improvement of existing ones (Mol & Birkinshaw, 2009; Clercq, Menguc, & Auh, 2009; Laforet, 2008).

Thus, this chapter contributes to the understanding of the effects that a wider cultural context can have on innovation. Does culture trigger the process of innovation? This chapter considers this issue by relating cultural dimensions with innovation indicators in the wider context of Europe. Then we look at the particular cases of Portugal and Spain in order to understand in more detail the effects of culture on innovation.

We start by framing the research and therefore present some studies on the relation between culture and innovation, especially research that used Hofstede's (1980, 2001) Cultural Dimensions Theory. The empirical analysis is divided in two parts: first we test the effect of culture on innovation using a log-linear econometric model, and then we move to the analysis of the Portuguese and Spanish cases. Finally, we summarize some conclusions.

CULTURE

There are as many definitions of culture as different cultures exist, probably because grasping all of cultural nuances and particularities is difficult. Tylor (1871) provided one of the first definitions of culture, and many of the definitions formerly used in social sciences present roughly the same dimensions. Trompenaars and Hampden-Turner (1998, 6) define culture as the "the way in which a group of people solves problems and reconciles dilemmas." Hofstede (1980; 2001) also defined culture as the collective mental programming that distinguishes members of a group and compared it to the "software of the mind," meaning that culture works as the social basis of human behaviour in a way that it can even influence the natural act of thinking (Hall, 1976).

One of the basic assumptions about culture is that it functions as a collective memory of the solutions a group or society developed to face everyday problems, based on their own understanding of the world (Schein, 1984), meaning that inventiveness is present, in some degree, in every group's culture. Defining innovation as "a complex and ongoing process of discovery, development, learning and application of new ideas, commercial and industrial methods as well as technologies" (Vieira, Neira, & Vazquez, 2008), we can say that innovation is, therefore, an important part of every culture.

According to several researchers (Trompenaars & Hampden-Turner, 1998; Schein, 2004) culture can be understood has having several layers. The outer layers are the most visible and changeable; the inner layers are the most hidden and unchangeable components of culture. The outer layers can be expressed in the material, concrete, and observable aspects of culture (such as food, dress, or language), while the inner aspects are the social and cultural values and norms that guide and give meaning to the material expressions of culture.

Culture can also be seen at different levels. National (or regional) level refers to the specific traits of a country's culture or a region, such as Western culture. Corporate culture refers to specific ways a group of people of the same organization share in solving problems (Schein, 2004). It is also possible to identify professional cultures, which are related to certain functions, such as doctors, lawyers, or marketing (Trompenaars & Hampden-Turner, 1998; Hofstede, 1980). Since our goal is to understand how culture can relate with countries' innovation, this research focuses on the national level.

How Can National Cultures Be Analysed?

The interest in national cultures has grown since the intensification of economic globalisation, with multinational companies facing different contexts, and the need of management to cope with different ways of thinking and acting. Thus, the study of national cultures has been developed on the basis of a comparative perspective. This means that, for example, no culture is absolutely individualist or collectivist; it means that culture A can be more or less individualist than culture B.

Several researchers developed models to analyse cultures. Trompenaars and Hampden-Turner (1998) developed a model with seven dimensions. This basic premise is that these dimensions are the basis of the way societies choose to cope with certain problems presented as dilemmas, namely, those which arise from our relationships with other people (the first five dimensions in Table 3.1.); those which come from the passage of time; and those which relate to the environment.

Table 3.1 Trompenaars & Hampden-Turner (1998) Cultural Dimensions

Dimensions	Description
Universalism vs. Particularism	Universalist cultures define what is good and right and apply it in any circumstances. Particularist cultures give more attention to unique circumstances. For example, in a particularist culture, friendship has particular obligations and therefore may come first.
Individualism vs. Communitarism	Refers to the focus on individual or community and what comes first. How people see themselves: as individuals or as part of group.
Neutral vs. Emotional	This dimension is related with the expression of emotions in relationships; that is, how far is it acceptable to demonstrate our feelings in different contexts.
Specific vs. Diffuse	Relates to the way people are involved in business relationships. A specific-type relationship is rational, objective, and "straight-to-the-point." On a diffuse-type relationship people engage in apparently useless activities not related with their primary objective.
Achievement vs. Ascription	Refers to how status is attributed. In an achievement-driven culture, people are judged on the basis of their accomplishments; in an ascription-driven culture people are judge by their attributes (e.g., age, gender, kinship, education).
Attitude to time	Refers to how societies look at time. Some cultures are more focused on the past and present; other cultures are more focused on the future.
Attitude to environment	Defines the relation people have with the surrounding environment. Some cultures see the surrounding world as more powerful than individuals; some cultures see their lives not as the result of the power of the environment but as the result of their own actions. It is the difference between affect and being affected by external factors.

Based on the premise that economic development is linked to values orientation, Inglehart (1997) analysed data from 43 countries collected for the 1990–1991 World Values Survey and found some patterns and cross-cultural differences. The results of the analysis revealed two main dimensions: the "tradition vs. secular-rational" dimension and the "survival vs. expression values" dimension. The dimension tradition is based on the finding of a certain regularity on a number of common characteristics. While traditional cultures tend to show an emphasis on male dominance in economic and political life, deference to parental authority, importance of family life, strong emphasis on religion and authority, among others, secular-rational societies tend to show the opposite characteristics. According to Inglehart, this opposition reflects the pre-industrial and industrial societies, respectively.

The dimension "survival vs. expression values" opposes materialist and post-materialist values, indicating a contrast between economic and physical security, on the one hand, and an emphasis on self-expression, subjective well-being and quality of life, on the other. Also, the expression values pole is identified with post-industrialist societies characterized by valuing trust, tolerance, subjective well-being, and self-expression. Inglehart considers that societies that tend to be more expression values oriented show more security in economic terms. On the other hand, societies with less security and low levels of well-being tend to emphasize economic and physical security above all. Other characteristic values of survival societies are low interpersonal trust, relative intolerance to outgroups, low support for gender equality, relatively high levels of faith in science and technology, and relative favourability to authoritarian governments.

Hofstede's Cultural Dimensions (Hofstede, 1980; 1997) is a very popular model. Hofstede's interest in cultural phenomena dates back to the 1970s when he started the study of cultural differences using IBM workers from more than 50 countries as an empirical ground. He starts from the definition of culture, which can be seen as the collective mental programming that identifies members of a group (Hofstede, 1997). This computer metaphor doesn't mean that there is no room for creativity; on the contrary, individuals can adapt their "software" in order to adjust to different contexts and goals. Another important point about culture is that it allows individuals and groups to solve problems, and, thus, facing the same problem, individuals from different cultures can present different solutions.

The theoretic model is made up of dimensions. In Hofstede's terms, this means that (1) they are independent of each other, (2) it is possible to combine them in different ways, and (3) they operate with two opposite extremes along a continuum. The theoretic model initially presented four dimensions and later a fifth dimension was added (Hofstede, 1983), as presented in Table 3.2.

Although Hofstede's Cultural Dimensions are a comprehensive model which allows the study of national cultures and the comparison between them, it has been subject to extensive criticism. One of its more fierce

Table 3.2 Hofstede Cultural Dimensions (1983; 1997)

Dimensions	Description
Power Distance (PDI)	PDI defines how people deal with inequalities. These inequalities can be measured in terms of power and wealth. The power distance index gives us a clue to the social and individual level of tolerance to those differences. This dimension seems to be correlated with collectivism: in countries where collectivism scores high, there is also a tendency to score high on power distance. However, the results are not so clear when applied to the relation between individualism and power distance.
Individualism (IDV)	This dimension refers to the relation between an individual and other individuals. At one end, there is individualism translated in very loose ties. This dimension seems to correlate with national wealth: more individualist societies tend to be wealthier.
Masculinity (MAS)	Masculinity accounts for the (social) division of roles between sexes. When a society is mainly "masculine" it means that masculine values – such as performance, achievement and materialism – spread out to all society, including women. The opposite, 'feminine' societies, are more concerned with relationships, quality of life, and environmental issues.
Uncertainty Avoidance (UAI)	Uncertainty avoidance refers to the way societies deal with the unknown, an unchangeable characteristic of the future. Societies that score low on uncertainty avoidance tend to prepare their members to accept uncertainty with ease so they are better equipped to take risks. Another characteristic of low uncertainty avoidance societies is the high level of tolerance regarding other people's opinions and behaviour.
Long/short term Orientation (LTO)	LTO deals with what has been called Virtue and Truth, which is found in the teachings of Confucius. The former is associated with thrift and perseverance; the latter emphasises tradition and the fulfilment of social obligations.

opponents is McSweeney (2002), who criticises the entire model, from the basis (the notion of culture) to the methodology approach. Others, such as Baskerville (2003), build their objections based on the argument that anthropology and sociology, that is, the scientific disciplines used to construct and refine the concept, do not apply to Hofstede's model.

It is not our goal to go through the arguments of McSweeney (2002) or Baskerville (2003) step by step, and Hofstede himself replied to these objections elsewhere (2002). However, it should be said that the model is far from perfect and, to cover all the aspects of such a complex concept as culture, attention should be brought to the wide applicability of its principles in areas such as organizations, consumption, tourism, marketing, and others. Furthermore, every theoretic development is subject to scrutiny, but it should be made on a constructive and not a destructive basis. In other words, the criticisms should be followed by a new enlightening alternative and this was not the case.

Innovation Indicators

Most of the literature pertinent to this subject seems to highlight the important role played by culture in enhancing a country's innovation. Thus, since we want this chapter to present an empirical analysis of the relationship between innovation and culture, it is vital to clarify the main methods used to quantitatively measure the efforts of innovation developed by a society.

Innovation can take on a number of dimensions within society, not only at an economic level but also, and increasingly, at a social and cultural level. Yet, these dimensions are difficult to measure, and, therefore, many of the attempts to measure innovation end up stemming from the economic dimension. The indicators most frequently used in the measurement of innovation are expenditure on Research and Development and the number of patents per worker (Mairesse & Mohen, 2003; Rao, Ahmad, Horsman, & Kaptein-Russell, 2001). Although we can indicate several studies that use these two innovation measures, we highlight the contribution of Mairesse and Mohnen (2003), which uses both variables to explain labour productivity, the first as an input variable and the second as an output one. However, it is possible to similarly highlight other indicators related to the adoption of new technologies and the level of skilled labour.

Investment in R&D seems to play an important role in generating new products at both the country and institutional levels. The expenditure on R&D is the main indicator used to reflect the efforts of innovation, and it functions as the input/catalyst of the whole innovative process. Despite its importance, this is not a perfect measure of innovation, especially since not all further research and development generates results. However, even though the ability to innovate is reduced, the ability to absorb technology developed by others is far greater in this case.

Another indicator that together with the expenditure on R&D is mentioned by scientific literature is the number of patents. This variable is often used as a product of innovation efforts and tends to measure a country's technological frontier. To explain OECD countries' labour productivity level, Rao et al. (2001) use patents and human capital variables which underline mostly the knowledge index of a country and enhance its cultural differences. Patents serve the innovators' interests, insofar as patents protect their intellectual property rights. This essentially represents a means of ensuring that innovators receive future returns of their investment in research and development and that these are not appropriated by imitators. This indicator is a measure of the output of the innovative process, although it only partially reflects it. Criscuolo and Haskel (2003) refer particularly to three factors: first, not all innovations reflect an invention, second, the patents have not commercial exploitation as its sole objective and, third, the propensity to patent inventions depends on the industry or area of work. As an alternative to patents, many institutions hold marketing secrets or use copyrights in order to protect their intellectual property rights. According to the model set up by Crepon, Duguet, and Mairesse (1998), the output of innovation is measured both by the number of patents as well as the percentage of sales derived from the innovative process (sales of new products to companies or the market).

However, the indicators of innovation are not limited to the two mentioned above. The adoption of new technologies is also an important indicator that reflects the impact of innovation on productivity; this may be measured by the number of new products and processes introduced and the percentage of output generated by these new products or processes. One of the sample measures regarding the adoption of new technologies is embodied in investment in machinery and equipment. Finally, the level of qualifications employees hold plays an important role in the measurement of innovation. This is supported by most scientific literature, in which we highlight Guisan and Aguayo (2005), who consider the level of skilled labour as an important input in the process of innovation. This can be translated by the percentage of employment of scientists, engineers, and other qualified professionals in R&D, over total employment.

In order to explain the effect of innovation on productivity, many studies use only one of these indicators. However, none of these indicators alone can reflect the multidimensional nature of the innovative process and the real level of a country or institution. Despite everything, it is possible to empirically acquire an approximate and more simplified perspective regarding the effect of innovation on productivity.

The Relation between Culture and Innovation

Several studies (see Table 3.3.) try to demonstrate the effects of culture on innovation using Hofstede's (1980) cultural dimensions. Shane (1992), using the concept of "inventiveness," concluded that individualism and power distance have an impact on the number of inventions patented per capita, but with a different reading: while individualism has a positive impact, power distance has a negative impact. Later, Shane (1993) studied

Table 3.3 Summary of Research on Innovation and Hofstede's Cultural Dimensions

Authors	Sample	Data Analysis	Dependent Variable	Independent Variables and Impact	
				Cultural Variables	Economic Variables
Shane (1992)	Countries: 33 Years: 1967, 1971, 1976, 1980	Rank correlations over split-sample methods of comparison	Per capita number of inventions patented	IDV + PDI -	
Shane (1003)	Countries: 33 Years: 1975–1980	Least squares multiple regression analysis	Per capita number of trademarks	IDV + UAI - MAS 0 PDI -	Per capita income + Industrial strucutre
Williams & McGuire (2005)	Countries: 63 Years: 1996, 1998, 1999, 2001	Structural equation model	National prosperity (output per worker; capital stock per worker; market capitalization % (GDP)	IDV + PDI - UAI -	Economic creativity (patents granted to residents; scientific and technical publications; R&D expenditure) + Innovation implementation (innovation implementation index; high technology exports) +
Vecchi & Brennan (2008)	Countries: 24 Years: 2006		Innovation inputs (technology adoption; R&D investment; management processes) Process innovation (product development procedures)	IDV - PDI + UAI - MAS - IDV - PDI + UAI 0 MAS -	

Culture Impact on Innovation 65

Authors	Sample	Data Analysis	Dependent Variable	Independent Variables and Impact		
				Cultural Variables	Economic Variables
Kaasa & Vadi (2010)	Countries: 20 at NUTS1 and NUTS2 level. Years: 2002 and 2004 (for cultural variables); 2003 (for patent applications)	Correlation analysis	Innovation initiation (patent applications) at NYTSI. Innovation initiation (patent applications) at NUTS2	IDV - PDI + UAI 0 MAS 0	
Halkos & Tzeremes (2011)	Countries: 25 Year: 2007	DEA estimator and bootstrap techniques for bias correction	Innovation efficiency: Innovation inputs (innovation drivers; knowledge creation; innovation & entrepreneurship). Innovation outputs (applications; intellectual property)	PDI - IDV - UAI - MAS +	

Note: "+" positive impact; "–" negative impact; "0" neutral or no impact.
IDV: individualism; PDI: power distance index; UAI: uncertainty avoidance index; MAS: masculinity.

Source: Authors.

innovation, measured by per capita trademarks, and completed the research using uncertainty avoidance and masculinity dimensions. The results for the impact of individualism and power distance were similar to his former research. Uncertainty avoidance revealed a negative impact on innovation and masculinity presented a neutral impact.

More recently, Williams and McGuire (2010) looked for the effects of culture on national innovation and prosperity. They divided innovation into two different phases: the initiation and the implementation phases, which they called "economic creativity" and "innovation implementation." The results show that culture does influence economic creativity and innovation implementation, which, in turn, has a positive effect on national prosperity. However, cultural dimensions did not behave in the same way: while individualism has a positive effect, uncertainty avoidance and power distance have a negative outcome.

In a study on international manufacturing, especially focused on ERP implementation, Vecchi and Brennan (2008) followed the multi-dimensional approach of Williams and McGuire (2010), and divided innovation into three different concepts: innovation inputs, innovation process, and overall innovation performance. The first two dimensions correspond to Williams and McGuire's (2010) economic creativity and innovation implementation, respectively. Contrary to previous research and all three concepts, power distance revealed a positive impact and individualism a negative impact. Uncertainty avoidance only showed a negative impact on the innovation inputs dimension and a neutral impact on innovation process and performance. The masculinity dimension also contradicted previous research, because it seems to have a negative impact on innovation inputs and processes.

Kaasa and Vadi (2010) developed a different approach. While emphasising the concept of innovation initiation through the use of patent applications indexes (for all patent applications, high-tech patent applications, information and communication technologies patent applications, and biotechnology sectors patent applications), they upgraded the data for the cultural dimensions using the European Social Survey indicators, and applied them to the NUTS1 and NUTS2 levels. When applying factor analysis, the study confirmed Hofstede's dimensions with the exception of individualism that revealed two factors (which were named overall individualism and family-related collectivism). A second-order factor analysis revealed two important factors: factor 1 grouped power distance, uncertainty avoidance, masculinity, and family related collectivism; factor 2 was made up of 'overall individualism.' The results showed that Factor 1 (power distance, uncertainty avoidance, masculinity, and family-related collectivism) has a negative impact on initiating innovation, especially uncertainty avoidance, and that overall individualism has a weaker or non-existent relationship with patenting intensity. In fact, overall individualism seems to have some impact on general patent applications at the NUTS2 level, but when looking at patent applications by sector and at the NUTS1 level, the results tend to be not significant.

Finally, Halkos and Tzeremes (2011) in a study of the effect of national culture on innovation efficiency with 25 countries, using conditional and unconditional Data Envelopment Analysis, found that higher power distance, higher individualism, and higher uncertainty avoidance had a negative impact on countries' innovation efficiency, whereas masculinity had a positive impact. Innovation was measured in terms of innovation input and output.

The results of the studies presented show that there is no consensus on the effect of culture on innovation. This does not mean that culture has no influence whatsoever on innovation. The differences can probably be found on the indicators used to measure innovation—which are slightly different in the studies presented–, but also on the approach to the concept of innovation. Some studies emphasized only innovation initiation stage, such as Shane (1992) and Kaasa and Vadi (2010); and others, such as Williams and McGuire (2010) and Vecchi and Brennan (2008) also measured the cultural effects on innovation implementation. Finally, it should be stressed that economic variables can also play an important role because, like Shane (1992; 1993) concluded, per capita income can have a positive effect on innovation.

METHODOLOGICAL GUIDELINES

According to previous work, the main goal of this chapter is to clarify the relation between innovation and culture based on the theoretical model of Hofstede's Cultural Dimensions. With the purpose of undertaking a balanced and sustained analysis of European countries, the sample used for the empirical study corresponds to 22 countries, analysed in a 15-year time period, between 1995 and 2009, which allowed a panel data estimation. The countries included in the sample are: Belgium, Bulgaria, Czech Republic, Denmark, Germany, Estonia, Ireland, Greece, Spain, France, Italy, Luxembourg, Hungary, Netherlands, Austria, Poland, Portugal, Romania, Finland, Sweden, United Kingdom, and Norway. Data availability was the main criterion for country selection. After the descriptive data analysis, we first studied the cointegration of the variables in the sample, in order to allow for the stationary panel data estimation.

Data is taken from Eurostat for the innovation variables (Table 3.4). We use R&D expenditure as a proxy for innovation. Data standardisation was made through the use of R&D expenditure per employee (in Euros, at 2000 constant prices) and the calculation of an index where the average for the countries studied is 100.

Cultural Dimensions data were taken from Hofstede (2001). The values presented are the result of Hofstede's studies from 1967 to 1972. and some of them were updated as the result of more recent studies. Following

Table 3.4 Variables Definition

Dependent Variable

GID_{it} — R&D expenditure, per employee (Euros, at 2000 constant prices) for the region i, year t.

Independent Variables

PRD_{it} — Gross Value Added, per employee (Euros, at 2000 constant prices) for the region i, year t.

PDI_{it} — Power Distance Index, for the region i, year t.

IDV_{it} — Individualism Index, for the region i, year t.

UAI_{it} — Uncertainty Avoidance Index, for the region i, year t.

MAS_{it} — Masculinity Index, for the region i, year t.

Note: Time period: 1995–2009.
Sources: Eurostat (for GID and PRD) and Hofstede, 2001 (for PDI, IDV, UAI, and MAS).

several studies (Williams & McQuire, 2005; Kaasa & Vadi, 2010) using Hofstede's Cultural Dimensions, it is assumed that the cultural characteristics remain relatively unchanged and, thus, can be represented as constant variables. Construct validity is assumed since it has been validated by many replication studies (Kogut & Singh, 1988; see Pagell et al., 2005, for an overview).

EMPIRICAL ANALYSIS

We start by describing the main variables used in the study. We choose to compare R&D expenditure with each of the cultural dimensions to favor the interpretation of the potential relation between innovation and culture.

DESCRIPTIVE ANALYSIS

There is a wide difference in the amount of expenditure in the selected countries (Figure 3.1). Nordic and Central European countries are among the countries that invest more in R&D. On the contrary, Eastern and Southern countries spend less. There is a general trend to increase the amounts of expenditure in almost every country, however with different paces. Countries such as Denmark, Ireland, Austria, and especially Finland show a significant increase in the amount of expenditure. On the contrary, countries such as France, Netherlands and Romania present a disinvestment trend in R&D. Finally, a small group of countries (Czech Republic, Estonia, Spain, and Portugal), although with expenditure levels above the average, present a significant investment effort that in some case represents the double or the triple invested in 1995.

Culture Impact on Innovation 69

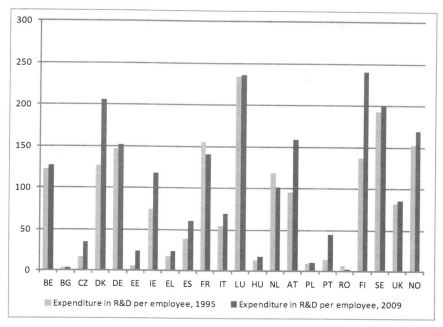

Figure 3.1 R&D expenditure per employee, 1995–2009 (100 = countries' average).

We now turn to the understanding of the relation between cultural dimensions and innovation. If we start the analysis with individualism (Figure 3.2), the presence of this characteristic is clear among Northern and Central European countries. According to the definition of individualism, these countries put more emphasis on personal freedom which can be an important feature to foster innovation. In fact almost all countries that score high on individualism are the ones that spend more in R&D per employee (right top quadrant), with the exception of the United Kingdom, Hungary, Italy, and Bulgaria. The right top quadrant—countries with higher scores on individualism and higher levels of R&D expenditure—includes countries mostly from the North and some from Central Europe, with a special emphasis on the Nordic group (Norway, Sweden, Denmark, and Finland). On the contrary, the left low quadrant reflects low scores on individualism and lower levels of R&D expenditure. Countries from Southern and Eastern Europe are well represented in this quadrant. Romania, Greece, and Portugal, but also Spain, Czech Republic, Estonia, and Poland are characterized by low individualism (or high collectivism) and low levels of R&D investment.

Masculinity presents a wide range of scores (Figure 3.3). The most masculine societies can be found in Austria, Italy, Ireland, Poland, United Kingdom, and Germany. This contrasts with the Nordic European countries,

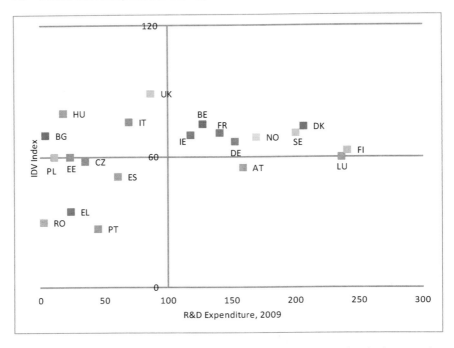

Figure 3.2 R&D expenditure per employee (2009) vs. individualism (index measure).

which are among the more feminine societies. Although there is not a clear trend, it is possible to group European countries into three different clusters: on the one hand, countries such as the Nordic that present low levels of masculinity but high levels of R&D expenditure; on the other hand, countries such as Hungary, Bulgaria, Estonia, Portugal, Spain, Romania, Greece, and Czech Republic with low levels of masculinity and also low levels of R&D expenditure. Another group with moderate scores on masculinity and high levels of R&D expenditure is made by Ireland, Austria and Germany, Belgium, and France.

Uncertainty avoidance index (UAI)—measuring the acceptance of risk taking with consequences for the need to develop strategies to control the future and making it predictable, namely through bureaucracy—follows the dispersion pattern of previous dimensions (Figure 3.3). However, it is possible to identify a very clear cluster of countries (top left quadrant) that present high levels of UAI and very low levels of R&D expenditure, including countries from the South and Eastern Europe, such as Greece, Portugal, Poland, Romania, Spain, Italy, and Czech Republic. Again, the Nordic European countries constitute the group of countries where R&D expenditure is high and UAI is low although with more outlying scores. Another group is made of Central European countries (Belgium, France, Austria,

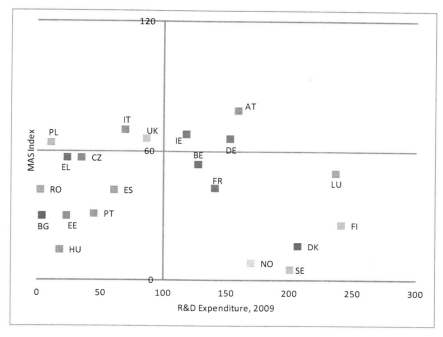

Figure 3.3 R&D expenditure per employee (2004) vs. masculinity (index measure).

and Germany) with moderate to high scores on UAI and also high levels of R&D expenditure. Estonia, Hungary, and Bulgaria constitute a group of countries that are characterized by relatively low UAI and low level of R&D expenditure. Finally the two Anglo-Saxon countries (UK and Ireland) present low scores for UAI and are near the average on R&D expenditure.

Finally, the Power Distance Index (PDI), which measures how societies deal with inequalities, follows the trends presented by other dimensions such as UAI (Figure 3.4). In this figure it is possible to identify two main groups: a very close group of countries (left quadrant) present low levels of R&D expenditure and moderate to high levels of PDI. This group is essentially made of Southern and Eastern European countries. On the other hand, there is a more dispersed group that presents low levels of PDI and high levels of R&D expenditure. This group is, once again, made of Nordic and Central European countries. Belgium and France contradict this grouping by presenting moderate levels of PDI, but R&D expenditure above the average.

This characterisation of the relative position of countries regarding R&D expenditure and the four Hofstede's Cultural Dimensions allows supporting two main ideas about the regional behaviour of countries. First, the Nordic countries systematically present the opposite profile of the Southern and Eastern European countries. While the former constitute the group

72 Pedro Ferreira, Elvira Vieira, and Isabel Neira

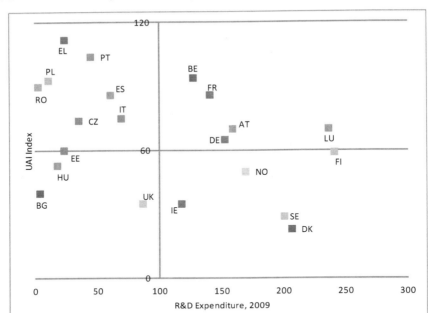

Figure 3.4 R&D expenditure per employee (2009) vs. Uncertainty Avoidance Index (index measure).

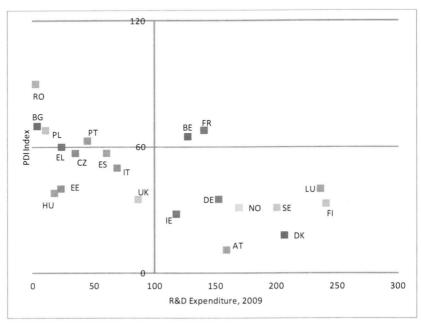

Figure 3.5 R&D expenditure per employee (2009) vs. Power Distance Index (index measure).

of heavy investors in R&D, the latter are well below the average. Second, while Nordic countries score high on individualism and low on masculinity, uncertainty avoidance, and power distance, the Southern and Eastern countries have an almost reverse cultural profile: moderate/low on individualism, low on masculinity, high on uncertainty avoidance, and moderate/high on power distance.

EFFECTS OF CULTURAL DIMENSIONS ON INNOVATION

In order to investigate the possible casual relation between culture and innovation (measured as an innovation input) we estimate a log-linear econometric model, in which R&D expenditure in European countries is explained from the following variables: gross value added per employee (productivity), Power Distance Index, Individualism Index, Uncertainty Avoidance Index, and Masculinity Index (see Table 3.5).

For the estimation, we have used an unbalanced pool equation because of the lack of data in some regions. The common coefficients cannot be accepted; in fact, on conducting the F test of common parametric stability, the model shows a lack of stability. We have taken into account the individual effects using estimations of Fixed Effects (FE) because of the results of the Hausman test, which shows that the estimator of the fixed effects is the only consistent one. The final estimation includes White cross-section standard errors and covariance.

Table 3.5 R&D Expenditure Econometric Model

Dependent Variable: Log(RD)							
Sample: 1995–2009 (15 obs.)							
Total Pool (unbalanced) observations: 287							
Estimation Method	Constant	Log(PRD)	Log(MAS)	Log(IDV)	Log(PDI)	Log(UAI)	R^2
Pooled LS	-8,07***	1,32***	-0.20***	0.49***	-0,29***	0,09	0.95

*, **, *** Significant at 1%, 5%, 10%, respectively

Table 3.6 Redundant Fixed Effects Tests

Equation: Untitled			
Test period fixed effects			
Effects Test	Statistic	d.f.	Prob.
Period F	0.743330	(14,267)	0.7296
Period Ch-square	10.973646	14	0.6881

The estimation FE using ordinary least squares, as well as the use of a redundancy test of fixed effects, leads to the acceptance of redundancy which means that the cultural variables collect fixed effects from the countries; therefore inexistent significant annual effects (see Table 3.6). All variables are significant in order to explain the R&D expenditure of European countries, except for the Uncertainty Avoidance Index.

In pooled estimation, we verified our hypothesis on the positive effect of the productivity on R&D expenditure, which means that this expenditure is even higher when the labour productivity rises and employees become more efficient. As for the cultural variables we find that they have a different behaviour when explaining the R&D expenditure. However this trend is partially consistent with Shane's study driven in 1993, based on Hofstede's cultural Indexes (1980). Shane (1993) found that Power Distant and Uncertainty Avoidance societies were less innovative while Masculine and Individualistic ones tend to encourage innovation.

In our study the conclusions are similar, although we register some differences, especially in the Masculinity Index estimator, which presents a negative value, contrary to Shane's findings. According to the study carried out by Shane (1993) masculine societies will be more innovative than feminine societies, whereas our study points to the trend of feminine societies toward being more innovative; to explain this discrepancy we have to consider several aspects. First, the time period is different, and cultural changes may have happened. Second, Shane (1993) used "per capita number of trademarks" as proxy which can be considered an output of innovation, while this study uses "R&D expenditure" which should be considered an input to the process of innovation. Finally, the uncertainty avoidance index estimator also presents a different behaviour, although we cannot make a solid comparison because of the lack of significance of this variable to the present model.

THE CASE OF PORTUGAL AND SPAIN

The analysis of the cultural dimensions in Spain and Portugal should be framed in the specific area of the countries analysed. In this sense, this section will discuss the four Hofstede dimensions in a comparative way relating them with other social indicators allowing a more complete picture of the social and cultural environment of both countries.

The Power Distance dimension deals with the fact that individuals in societies are not equal; it expresses the attitude of one's culture towards these inequalities. Spain's (57) and Portugal's (63) scores on this dimension reflects that hierarchical distance is well accepted and those holding the most powerful positions are assumed to have privileges for their position. According to Hofstede (2001), this dimension implies the existence of pronounced hierarchical structures.

Culture Impact on Innovation 75

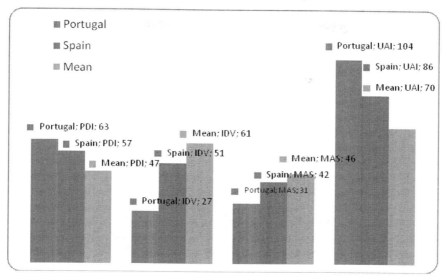

Figure 3.6 Cultural dimensions: Portugal, Spain, and European countries' average.

Trust in institutions, one of the more used indicators of social capital, can be related to the concept of power distance. According to data from the European Social Survey (ESS) in 2010, in terms of political trust, Portugal is one of the countries with less trust, presenting values less than 13% (percentage of people who claim to trust in institutions) and a score with a value less than 3 (out of 10). Spain is closer to the EU average, but less than 30% with a mean value of 4,29. A similar situation is observed in the confidence in politicians. Portugal is just over 5% and an average less than 2; Spain is the fourth country with the lowest confidence level (both in the mean and percentage values). Thus, and in general, Portugal and Spain are societies with a low level of generalized trust which has been shown to influence economic development.

Uncertainty Avoidance has to do with the way that a society deals with the fact that the future can never be known. The reaction to this universal fact and how societies deal with it describes this cultural dimension. Spain (with 86 points) and especially Portugal (with a score of 104, surpassed only by Greece), show a very high preference for avoiding uncertainty. Risk aversion is one of the factors discussed in the literature on entrepreneurship. Reducing the perception of failure increases the likelihood that an individual decides to start a new business (Weber & Milliman, 1997) with entrepreneurial behaviour being associated with moderate levels of risk taking (McCelland, 1961; Sexton and Bowman, 1983). When looking at the 2010 Total early-stage Entrepreneurial Activity (TEA) rate, an indicator of the Global Entrepreneurship Monitor survey, Spain with 4,3% and

Portugal with 4,5% present the lowest scores of all the OCDE countries. These are therefore two cultures with low rates of risk taking, where starting a new business is a scenario for very few individuals when compared with other European countries, where the TEA is more than 7% (which is the case of Latvia, Netherlands, or Norway).

Also, according to the Global Competitiveness Report of the World Economic Forum (2011), "restrictive labor regulations" and "inefficient government bureaucracy" are the second and third most problematic factors for doing business in Portugal and Spain. This illustrates the need that Portuguese and Spanish societies have in controlling the future and making everything predictable, in order to reduce the collective anxiety resulting from the high uncertainty aversion.

Individualism is the cultural dimension that reflects the concern about oneself and the immediate family only. In this context, the value of the individual is framed by loosely social ties. The opposite of individualism is collectivism (represented by the low end of the scale) where individuals belonging to a group are supposed to take care of one another. Thus, social ties tend to be stronger, and culture emphasizes the "we."

Portuguese (27) and Spanish (57) cultures are more collectivist than the majority of European countries, meaning that in these two societies, the comfort of group protection is well appreciated. In fact, when we look at the type of families, according to the Eurostat, Portugal and Spain are among the countries where the number of extended families is more persistent (8,82% and 9,94%, respectively) when compared with the rest of Europe, especially more individualistic countries such as the UK (4,16%) or the Netherlands (1,39%).

However when we look at some data related with organizations, the picture seems to be quite different. For example, teamwork is seen as an expression of collectivist cultures, since it reinforces tight social ties and group cohesiveness. According to data from the European Working Conditions Survey, fewer than half of Portuguese (49%) and Spanish (43%) workers are involved in any form of teamwork, well below the European average (64%). Thus, we might say that the collectivism that characterizes Portugal and Spain maybe of an informal type, which is supported by the "family-related collectivism" concept advanced by the findings of Kaasa and Vasi (2010).

Masculinity, and its counterpart, femininity, refers to how a society defines success. A masculine society will value achievement and competition, while a feminine society will value harmony, caring for others, and quality of life. In fact, feminine cultures define success by the degree of quality of life. Individuals in masculine cultures strive to stand out, competing and comparing with others, facing challenges as something to be won. On the contrary, standing out from the crowd is not valued or admired in feminine cultures; harmony and consensus are the keywords. Masculine cultures can be more conflicting. Portugal (31) and Spain (42) are more feminine than masculine. This means that these two countries prefer harmony and consensus rather than conflict and competition. Moreover, we are dealing with two cultures that value the quality of life, which is often related to leisure and the relatively less importance given to work.

The results of the previous section show that culture has an impact on innovation, namely, at the input level. Cultures characterised by individualism, low power distance, and feminine rather than masculine offer a more favourable context for innovation initiation. Thus, it is not surprising that Spain and Portugal present lower levels of innovation when compared with their European counterparts. In fact, apart from minor differences, Portuguese and Spanish cultures accept well inequalities and hierarchical differences (high power distance), avoid conflicts and competition, preferring harmony (low on masculinity), which is reinforced by the value of the group over the individual (low on individualism), especially in informal contexts. Finally, and most important, Portugal and Spain stand out for the high level of uncertainty avoidance, only surpassed by Greece. With the exception of the feminine trait, the cultural context is not very favourable to innovation in both countries.

CONCLUSIONS

This study had two goals. On the one hand, following a macro perspective, we aim to analyse the effects that culture has on innovation, mainly the innovation initiation. On the other hand, based on the empirical findings of the first goal, we look closer at two of the Southern European countries, their cultural specificities, and how they could be related with the innovation context.

The first major idea is that the differences between European countries are pronounced. In regional terms, Nordic and Central European countries seem to have similar cultural characteristics, which are followed by similarities in innovation. The same can be applied to Southern and Eastern European countries, however, for the opposite reason, meaning that countries from these two regions have low investment in R&D. Second, culture does have a clear impact on innovation initiation: lower power distance, feminine, and individualistic cultures seem to foster innovation.

The outlook of Portugal and Spain was then examined in more detail. Being among the less innovative European countries, their cultural profile is also very similar. With a propensity for feminine values, high levels of power distance acceptance, and markedly collectivist traits, Portugal and Spain present a cultural environment with not so encouraging prospects as far as innovation is concerned.

Some limitations can be pointed to this research. Although it is possible to state that Hofstede's Cultural Dimensions are a useful tool in cross-cultural studies, we should recognize that, despite the stability that characterises the cultural phenomenon, up-to-date data on cultural dimensions for the period in study would be more accurate. Moreover, we analysed four of the five dimensions proposed by Hofstede. Long-term orientation is relatively new, and data are from a different time frame. On the other hand, this study is based only on European countries and the conclusions cannot be generalized to other regions. Finally, we must recognise that innovation is a complex concept and difficult to grasp in a single indicator. Nevertheless, R&D reflects a political and managerial option to initiate the process of innovation, and as it was shown, culture plays an important role in shaping the way different options are considered.

REFERENCES

Baskerville, R. (2003) Hofstede never studied culture. *Accounting. Organization and Society*; 28, 1–14.
Breitung, J. (2002) Nonparametric test for unit roots and cointegration. *Journal of Econometrics*, 108, 343–363.
Clercq, D., Menguc B., & Auh S. (2009) Unpacking the relationship between an innovation strategy and firm performance: The role of task conflict and political activity. *Journal of Business Research;* 82(November), 1046–1053.
Crepon, B., Duguet E., & Mairesse J. (1998) Research, innovation and productivity: an econometric analysis at the firm level. NBER Working Paper 6696.
Criscuolo, C., & Haskel J. (2003) *Innovations and productivity growth in the UK: Evidence from CIS 2 and CIS3*. London: Center for Research into Business Activity (CeRiBA).
Fisher, R.A. (1932) *Statistical methods for research workers*, 4th ed. Edinburgh: Oliver & Boyd.
Guisan, M., & Aguayo E. (2005) Employment, development and research expenditure in the European Union: Analysis of causality and comparison with the United States, 1993–2003. *International Journal of Applied Econometrics and Quantitative Studies*; 2 (2), 21–30.
Hadri, K. (2000) Testing for stationary in heterogeneous panel data. *Econometrics Journal*, 3, 148–161.
Halkos, G.E., & Tzeremes, N.G. (2011) The effect of national culture on countries' innovation efficiency. MPRA Paper 30100. *http://mpra.ub.uni-muenchen.de/30100*. Accessed January, 2012.
Hall, E.T. (1976) *Beyond culture*. Garden City: Anchor Press.
Hofstede, G. (1980) *Culture's consequences*. Beverly Hills, CA: Sage.
Hofstede, G. (1983) The cultural relativity of organizational practices and theories. *Journal of International Business Studies*, 14, 75–89.
Hofstede, G. (1997) *Culturas e organizações. Compreender a nossa programação mental*. Lisbon: Edições Sílabo.
Hofstede, G. (2001) *Culture's consequences: Comparing values, behaviours, institutions and organizations across countries*. Thousands Oaks, CA: Sage.
Hofstede, G. (2002) Dimensions do not exist. A reply to Brendan McSweeney. *Human Relations*; 55 (11), 1355–1361.
Inglehart, R. (1997) *Modernization and Postmodernization. Cultural, economic and political change in 43 societies*. New Jersey: Princeton University Press.
Im K., Pesaran M., & Shin Y. (2003) Testing for unit roots in heterogeneous panels. *Journal of Econometrics*, 115, 53–74.

Kaasa, A., & Vadi M. (2010) How does culture contribute to innovation? Evidence from European countries. *Economics of Innovation and New Technology*; 19 (7), 583–604.
Kogut, B., & Singh H. (1988) The effect of national culture of the choice of entry mode. *Journal of International Business Studies*, 19, 411–430.
Laforet, S. (2008) Size, strategic, and market orientation affects on innovation. *Journal of Business Research*, 61 (July), 753–764.
Levin, A., Lin, C. & James Chu, C. (2002) Unit root test in panel data: asymptotic and finite-sample properties. *Journal of Econometrics*, 108, 1–24.
Maddala, G.S., & Wu, S. (1999) A comparative study of unit root tests with panel data and new simple test. *Oxford Bulletin of Economic and Statistics*, 61 (S1), 631–659.
Mairesse, J., & Mohen, P. (2003) R&D and productivity: A reexamination in light of the innovation surveys. DRUID Summer Conference. Copenhagen, 12th—14th June.
McClelland, D.C. (1961). *The achieving society*. Princeton: Van Nostrand.
McSweeney, B. (2002) Hofstede's model of national cultural differences and their consequences: A triumph of faith—a failure of analysis. *Human Relations*, 55 (1), 89–118.
Mol, M.J., & Birkinshaw, J. (2009) The sources of management innovation: When firms introduce new management practices. *Journal of Business Research*, 62 (December), 1269–1280.
Pagell, M., Katz J.P., & Sheu, C. (2005) The importance of national culture in operations management research. *International Journal of Operations Production Management*, 25 (4), 371–394.
Rao, S., Ahmad A., Horsman W., & Kaptein-Russell P. (2001) The importance of innovation for productivity. *International Productivity Monitor*, 2, 11–18.
Schein, E. (1984) Coming to a new awareness of organizational culture. *Sloan Management Review*, 25 (2), 3–16.
Schein, E. (2004) *Organizational culture & leadership*, 3rd ed. San Francisco: Jossey-Bass.
Sexton, D.L., & Bowman, N. (1983). Determining entrepreneurial potential of students. *Academy of Management Proceedings*, 42, 408–412.
Shane, S. (1992) Why do some societies invent more than others?. *Journal of Business Venturing*, 7, 29–46.
Shane, S. (1993) Cultural influences on national rates of innovation. *Journal of Business Venturing*, 8, 59–73.
Trompenaars, F. & Hampten-Turner, C. (1998) *Riding the waves of culture: Understanding diversity in global business*, 2nd ed. London: Nicholas Brealey.
Tylor, E. (1871) *Primitive Culture. Researches into the development of mythology, philosophy, religion, art, and custom*, Vol. I. London: John Murray
Vecchi, A., & Brennan, L. (2008) A cultural perspective on innovation in international manufacturing. *Research in International Business and Finance*, 23(2), 181–192.http://www.sciencedirect.com/science/article/pii/S027553190800024X. Accessed January, 2012.
Vieira, E., Neira I., *Vazquez E. (2008) Productivity and innovation economy: Comparative analysis of European NUTS II (1995-2004). *The Annual International Conference—Regions: The Dilemmas of Integration and Competition*. Prague: University of Economics.
Weber, E.U., & Milliman, R.A.(1997) Perceived risk attitudes: Relating risk perception to risky choice. *Management Science*, 43, 123–144.
Williams, L.K., & McGuire, S.J.J. (2010) Economic creativity and innovation implementation: The entrepreneurial drivers of growth? Evidence from 63 countries. *Small Business Economics*, 34 (4), 391–412.
World Economic Forum. (2011) *The global competitiveness report, 2011-2012*. Geneva: World Economic Forum.

4 Communities as Spaces of Innovation

Joanne Roberts

INTRODUCTION

In the contemporary economic environment, the competitiveness of firms and nations depends increasingly on the ability to sustain continuous and speedy innovation by harnessing sources of creativity and developing new knowledge. The need to organize for creativity and innovation has driven the development of a range of flexible and fluid organizational forms from the virtual or boundaryless organization to the shamrock or cellular form, from the intelligent enterprise and the project organization to the spaghetti organization and the hypermodern organization (see, for example, Handy, 1998; Miles, Snow, Mathews, Miles, & Coleman, 1997; Quinn, 1992; DeFillippi, 2002; Foss, 2006; Roberts & Armitage, 2006). It is within this context that community has attracted much attention among management scholars and practitioners as an alternative or complementary organizational form (see, for example, Wenger & Snyder, 2000; Wenger, McDermott, & Snyder, 2002; Mintzberg, 2009).

In addition, the role of communities in the process of knowledge transfer and generation has been the subject of research in various inter- and extra-organizational contexts from epistemic communities of policymakers and scientists to professional organizations and interest groups (Haas, 1992; Knorr-Cetina, 1999; Benner, 2003; Faulconbridge, 2010; Hall & Graham, 2004, inter alia). The communities of practice framework has been widely used to analyse the learning and knowledge generation activity occurring within such communities. Indeed, communities of practice, which have been identified as mechanisms through which knowledge is held, transferred, and created (Brown & Duguid, 1991; Lave & Wenger, 1991; Orr, 1996; Wenger 1998, 2000), have been increasingly used by management academics and practitioners, as a means of analysing and facilitating knowledge transfer and generation.

A community-based perspective complements established approaches to understanding and promoting the creation of knowledge and innovation in spatially specific contexts whether at an urban, regional, or national level: including, for instance, national and regional systems of innovation

(Lundvall, 1992; Nelson, 1993; Howells, 1999, ; inter alia), clusters (Porter, 1998; Malmberg & Maskell, 2002; Bresnahon & Gambardella, 2004, *inter alia*), industrial districts (Becattini, 1990, 2004; Markusen, 1996), innovative milieus (Camagni, 1995), and networks (Håkansson, 1987, Bjorg & Isksen, 1997). Many of these approaches focus on the macro dimensions of innovation, such as institutional and policy structures. In contrast, a focus on community offers the potential to explore the social interactions that underpin knowledge creation and innovation at a micro level (Gertler, 2008; Cohendet & Simon, 2008; Gertner, Roberts, & Charles, 2011). Hence, communities of practice are of interest to social scientists and policy makers concerned with innovation. It is then no surprise to find that the notion of communities of practice is receiving increasing attention from academics in the fields of economics and economic geography (see, for example, Amin & Cohendet, 2004; Coe & Bunnell, 2003; Benner, 2003; Amin & Roberts, 2008a).

The aim of this chapter is to assess the role of community as a facilitator of innovation. Acknowledging the place of community in innovation allows the social dimensions to be incorporated into understandings of knowledge generation. Moreover, the inclusion of a social dimension also provides scope for insights into the spatial dynamics of innovation. It is important for differentiating between communities of practice in terms of both their spatial reach and their social dynamics. Through a consideration of these aspects it will be possible to assess the benefits and limitations of communities of practice as spaces of innovation in a variety of social contexts.

The chapter begins by exploring the turn to community in contemporary economic thought before exploring early conceptualizations of communities of practice (Lave and Wenger, 1991; Wenger, 1998). Attention then focuses on the size and spatial reach of communities of practice and the implications of these qualities for their social dimensions. The challenges and opportunities offered by communities of practice as spaces of innovation are then outlined. The chapter concludes with a brief assessment of the place of communities of practice among approaches to understanding innovation at various spatial scales together with tentative policy suggestions to promote the role of community as a facilitator of innovation.

THE TURN TO COMMUNITY

The idea of community is the subject of extensive scholarly research in the fields of sociology, political science, and economics (see, for example: Tonnies, 1963; Etzioni, 1996; Putnam, 2000; Bowles & Gintis, 2002). The nebulous nature of the term community has resulted in a variety of definitions (Etzioni, 2000). For instance, Bowles and Gintis define a community as:

a group of people who interact directly, frequently and in multi-faceted ways. People who work together are usually communities in this sense, as are some neighbourhoods, groups of friends, professional and business networks, gangs, and sports leagues. The list suggests that connection, not affection, is the defining characteristic of a community. Whether one is born into a community or one entered by choice, there are normally significant costs to moving from one to another (Bowles and Gintis, 2002, 420).

Defined in this way, communities can be found everywhere, that is, wherever people interact and develop relationships. As Bowles and Gintis (2002) note, communities draw upon the various mechanisms that people have traditionally used to regulate their common activity, including trust, solidarity, reciprocity, reputation, personal pride, and respect, together with mechanisms to punish nonreciprocal behaviour including vengeance and retribution.

For Etzioni (1996) the relationships that make up community are viewed as affect-laden rather than merely connections. Community is:

a combination of two elements: A) A web of affect-laden relationships among a group of individuals, relationships that often crisscross and reinforce one another (rather than merely one-on-one or chainlike individual relationships). B) A measure of commitment to a set of shared values, norms, and meanings, and a shared history and identity—in short, to a particular culture. (Etzioni, 1996, 127)

Here Etzioni (1996) implies that a community is not necessarily locationally bound since direct interaction is not a defining characteristic. Consequently, such a definition is of particular value when exploring spatially distributed communities. Moreover, the reference to crisscrossing relationships resonates with the online network type communities that have flourished since the rise of the Internet.

For the purposes of this chapter, and drawing on the above definitions, *community is defined as characterized by relationships among a group of people, which are governed by an accepted set of norms and values.* Such communities are not new, but have existed since the dawn of human society. Equally, community is not a new phenomenon in the business context (Storper 2008), yet it does tend to be overshadowed by markets and hierarchies, which are the dominant modes of organizing economic activity. Nevertheless, in recent years, there has been a resurgence of interest in community as an alternative or complementary organizational form (Amin & Roberts, 2008a). Indeed, in the wake of the 2008 financial crisis, Mintzberg (2009, 7) calls for companies to rebuild as communities, noting that 'some of the companies we admire most—Toyota, Semco (Brazil), Mondragon (a Basque federation of cooperatives), Pixar, and so on—typically have this strong sense of community.'

The governance of communities depends on social capital which Putman (2000, 19) defines as 'connections among individuals—social networks and the norms of reciprocity and trustworthiness that arise from them.' The presence of social capital can improve the efficiency of organizations by facilitating the coordinated actions of individuals. Indeed, Adler (2001) forwards community as a third organizational form relative to markets and hierarchy, arguing that with its reliance on trust as its key coordination mechanism, rather than price in the market and authority in the hierarchy, community has a stronger capacity to effectively manage knowledge assets. Consequently, at the level of business organizations, community has become an attractive alternative in economies increasingly dependent on knowledge-based assets, which are often embedded in a creative workforce. Monitoring high skilled knowledge workers is difficult and sometimes impossible, thereby undermining any recourse to contractual and hierarchical management arrangements. Hence, rather than aligning workers' interests with the organization through the use of incentives and the exertion of authority, promoting a community-based organization allows alignment to occur through the adoption and promotion of certain social norms.

To secure competitiveness, based on a steady flow of creative products and processes, companies, large and small, are seeking to harness the innovative resources emerging in social environments and through local and distributed communities of all sorts. Although communities may be seen as supplementary to formal organizations, whether they are situated within, between, or beyond the boundaries of firms, they are not synonymous with them. Communities that reach across organizations give firms opportunities to access knowledge from the wider environment. Such communities include the social and professional communities in which employees participate. By maintaining links to external communities, such as professional bodies and trade associations, business organizations maintain open channels through which external knowledge can be accessed. Interaction with such communities adds to the porosity of the organization's boundaries and thereby to its flexibility and ability to respond to changing market opportunities and challenges. Social interaction is a powerful source of creative stimulus; free from the imperatives of ownership and control, ideas can flow freely among community participants (Roberts, 2010a). As a vehicle or container of sociality, community is therefore fertile ground for creativity that can be harnessed for commercial innovative activity (Leadbeater, 2008).

Communities and communities of practice are fairly synonymous, the key point of departure being the centrality of a specific practice in the latter. All communities engage in a variety of practices, whereas communities of practice form around specific practices. Nevertheless, communities of practice cover a wide range of activities. The idea of communities of practice is therefore malleable and can be applied in a wide variety of contexts. This quality in part accounts for the notion being taken up by management

COMMUNITIES OF PRACTICE

The concept of communities of practice was originally developed by Lave and Wenger (1991) in a study of situated learning in the context of five apprenticeships: Yucatec midwives; Vai and Gola tailors; naval quartermasters; meat cutters; and non-drinking alcoholics. Lave and Wenger (1991, 98) argue that a community of practice, which they define as 'a system of relationships between people, activities, and the world; developing with time, and in relation to other tangential and overlapping communities of practice,' is an intrinsic condition of the existence of knowledge. The communities of practice approach focuses on the social interactive dimensions of situated learning, a subject that has received attention from a variety of organizational researchers (see, for example, Orr, 1996; Barley and Orr, 1997; Blackler 1995; Boland & Tenkasi, 1995; Gherardi et al., 1998; Carlile, 2002). Brown & Duguid, (1991, 1998) introduced the notion of communities of practice to the field of management.

Through a study of an insurance claims processing office, Wenger (1998) developed a detailed understanding of the social dynamics of communities of practice. He argues that communities of practice are important places of negotiation, learning, meaning, and identity. For instance, within communities of practice, meaning is negotiated through a process of participation and reification. 'Any community of practice produces abstractions, tools, symbols, stories, terms, and concepts that reify something of that practice in a congealed form' (Wenger 1998, 59). Such forms take on a life of their own outside their original context where their meaning can evolve or even disappear.

Wenger (1998, 72–84) identifies three dimensions of the relation by which practice is the source of coherence of a community. First, members interact with one another, establishing norms and relationships through *mutual engagement*. Second, members are bound together by an understanding of a sense of *joint enterprise*. Finally, members produce over time a *shared repertoire* of communal resources, including, for example, language, routines, artefacts, and stories. Furthermore, Wenger (2000, 227–228) distinguishes between three modes of belonging to social learning systems. First, *engagement* is achieved through doing things together, for example, talking and producing artefacts. Second, *imagination* involves constructing an image of ourselves, of our communities, and of the world, in order to orient ourselves, to reflect on our situation, and to explore possibilities. Finally, *alignment* involves making sure that our local activities are sufficiently aligned with other processes so that they can be effective beyond our own engagement.

The existence of a community of practice may not be evident to its members because, as Wenger (1998, 125) notes, 'a community of practice need not be reified as such in the discourse of its participants.' Nevertheless, he argues that a community of practice does display a number of characteristics including those listed in Table 4.1.

Communities of practice are not stable or static entities. They evolve over time as new members join and others leave (Borzillo, Aznar, & Schmitt, 2011). Communities of practice as defined by Lave and Wenger (1991) cannot be formed. For example, a business can establish a team for a particular project, which may, in time, emerge as a community of practice. But management cannot establish a community of practice. What it can do is facilitate the spontaneous emergence of communities of practice and support those communities of practice that do develop. As Brown and Duguid (2001a) suggest, managers can seek to structure spontaneity; in particular, they have a role to play structuring fragmented practice across their organization. On the one hand, managers have a role supporting the development of communities of practice. On the other, they can encourage alignments of changing practices between communities, thereby assisting the transfer of knowledge across the organization (Brown and Duguid, 2001a).

Managers are seeking to develop and support communities of practice as part of their knowledge management strategies and, in this sense,

Table 4.1 The Characteristics of Communities of Practice

Sustained mutual relationships—harmonious or conflictual
Shared ways of engaging in doing things together
The rapid flow of information and propagation of innovation
Absence of introductory preambles, as if conversations and interactions were merely the continuation of an ongoing process
Very quick setup of a problem to be discussed
Substantial overlap in participants' descriptions of who belongs
Knowing what others know, what they can do, and how they can contribute to an enterprise
Mutually defining identities
The ability to assess the appropriateness of actions and products
Specific tools, representations, and other artefacts
Local lore, shared stories, inside jokes, knowing laughter
Jargon and shortcuts to communication as well as the ease of producing new ones
Certain styles recognized as displaying membership
A shared discourse reflecting a certain perspective on the world

Source: Compiled from Wenger (1998, 125–126).

communities of practice can be viewed as a new organizational form (Wenger & Snyder, 2000, Wenger, McDermott, & Synder, 2002), which can create value and improve performance (Lesser & Storck 2001). Furthermore, Swan, Scarbrough, and Robertson (2002) suggest that the notion of communities of practice can be used as a rhetorical tool to facilitate the control of professional groups over which managers have little authority. Additionally, much research concerning the transfer of knowledge and information in virtual organizations has been influenced by the communities of practice literature (see, for example, Johnson, 2001; Pan & Leidner, 2003; Hara, Shachaf, & Stoerger, 2009).

Communities of practice may be a part of a number of constellations of communities of practice sharing a variety of characteristics (Table 4.2.). According to Wenger (1998) when a social configuration is viewed as a constellation rather than a community of practice, the sustaining of the constellation must be maintained in terms of interactions among practices involving boundary processes. Wenger (1998, 2000) identifies a number of boundary processes through which knowledge can be transferred including brokering, boundary objects (Star & Griesemer, 1989), boundary interactions, and cross-disciplinary projects (Table 4.3.). Elements of styles and discourses can travel across boundaries (Wenger, 1998, 129), diffusing through constellations they can be shared by multiple practices and creating forms of continuity that take on a global character. For instance, Roberts (2010b) employs the notions of communities and constellations of practice to trace the diffusion and creation of management knowledge occurring through boundary processes involving management gurus, management consultants, business schools/management academics, managers, and business media.

Table 4.2 Characteristics of Constellations of Practice

Sharing historical roots
Having related enterprises
Serving a cause or belonging to an institution
Facing similar conditions
Having members in common
Sharing artefacts
Having geographical relations of proximity or interaction
Having overlapping styles or discourses
Competing for the same resources
May or may not be named/identified by participants
May or may not have people endeavouring to keep it together
May be created intentionally by boundary straddling individuals
May emerge from circumstances

Source: Compiled from Wenger (1998, 126–168).

Table 4.3 Types of Boundary Processes

Brokering	Boundary spanners: taking care of one specific boundary over time. Roamers: going from place to place, creating connections, moving knowledge. Outposts: bringing back news from the forefront, exploring new territories. Pairs: personal relationships between two people.
Boundary Objects	Artefacts: tools, documents, or models. Discourses: common language. Processes: explicit routines and procedures.
Boundary Interactions	Boundary encounters: visits, discussions, and sabbaticals. Boundary practices: practices developed to facilitate the effective crossing of boundaries. Peripheries: facilities by which outsiders can connect with practice in peripheral ways—'frequently asked questions,' introductory events, and 'help desks.'
Cross-disciplinary Projects	Cross-functional projects and teams that combine the knowledge of multiple practices to get something done.

Source: Compiled from Wenger (2000, 235–238).

SIZE AND SPATIAL RESEARCH

Communities of practice were originally presented as spontaneous, self-organizing, and fluid processes (Lave and Wenger, 1991). However, in later work Wenger and Snyder (2000), among others (Wenger et al., 2002; Saint-Onge and Wallace, 2003), suggest that they can be cultivated and leveraged for strategic advantage and be applied in a wide variety of intra-, inter-, and extra organizational contexts. For instance, Wenger et al. (2002), consider communities of practice in large multinational organizations, including Shell Oil Company, Daimler Chrysler, Hewlett Packard Company, McKinsey and Company, and the World Bank. Not only are communities of practice applied in large multinational organizations but some are also identified as having very large memberships. For example, Shell Exploration and Product International Ventures includes a globally distributed community of more than 1,500 members (Wenger et al., 2002, 115). Indeed, online communities, like that which has formed around Wikipedia with some 100,000 regular contributors,[1] have been identified as communities of practice (O'Sullivan, 2009). While it may be possible to identify communities of practice in both small groups of people working in close proximity and in globally distributed communities of 100,000 people, there are significant differences between these two types of communities of practice in terms of the nature of social interaction.

In some senses, some large distributed communities can be viewed as constellations of practice, which arise from interactions among practices involving boundary processes. The boundaries between communities of practice are not fixed, but flexible, continuously shifting, porous in nature, and difficult to identify. Although communities may originate in a local context, sustained and repeated interaction facilitated by various boundary processes may create new spatially extensive communities and constellations (Coe and Bunnell, 2003, 446). Technological developments in transportation and information and communication technologies (ICTs) are increasing the scope of engagement, but, as Wenger (1998, 131) argues, these developments involve trade-offs that reduce participation in the complexity of situations and their local meanings.

However, Amin (2002) suggests that organizational or relational proximity, achieved through communities of practice, may in reality be more important than geographical proximity. Relational proximity, usually achieved through face-to-face interaction, may also be achieved through ICTs and the mobility of individuals (Coe and Bunnell, 2003, 445). Indeed, Sole and Edmondson (2002, 32), noting the importance of the mobility of people in multi-site teams, claim that 'dispersed teams may be successful ... because they have enhanced awareness of a greater breadth of situated knowledge from which they are ... better positioned to learn.' Hence, distributed communities may hold greater scope for innovative activity than those that are co-located.

Traditionally economic geographers have viewed innovation as locationally bound because of its dependence on tacit knowledge embedded in spatially specific actors. However, they have begun to recognize how local and global actors may interact and contribute to innovation (Gertler & Levitte, 2005; Boschma, 2005; Bathelt, Malmberg, & Maskell, 2004; inter alia). Spatially distributed communities of practice provide a mechanism through which globally distributed knowledge can contribute to locally specific innovation projects.

The notion of constellations of practice, together with other concepts such as fractal structures for global communities (Wenger et al., 2002, 127), helps to incorporate spatially dispersed, virtual, and distributed communities as well as very large communities into a communities-of-practice framework. However, Brown and Duguid (1991) argue that all but the smallest of organizations should be regarded as communities of communities of practice. They make use of the term networks of practice to describe relations among members which are significantly looser than those in a community of practice (Brown & Duguid, 2001b, 205). While members of such a network are able to share knowledge, most of them will never know or meet one another. Yet, in her study of communities of practice in a high-technology firm, Teigland (2000, p. 143) found that Internet communities exhibit many of the characteristics of communities of practice even though the individuals involved never meet.

Extra-organizational communities of practice exist at various spatial scales. They have, for instance, been identified in studies of urban economic development in relation to silicon alley in Newcastle upon Tyne (Conway, Dawley, & Charles, 2005) and as facilitators of creativity in the city of Montréal (Cohendet, Grandadam, & Simon, 2010). Benner (2003) examines the value of communities of practice in supporting individual learning and collective learning processes in Silicon Valley using a case study of an association of women in Internet design and development occupations. In addition, Saxenian's (2006) study of Silicon Valley's immigrant high-technology entrepreneurs provides an excellent illustration of how community ties can facilitate learning and the construction of across-borders communities that reach from the Silicon Valley in the United States to Israel's Tel Aviv and Taiwan's Hsinchu regions and to China's Beijing and Shanghai regions and India's Bangalore and Hyderabad regions.

The film industry provides an exemplary case of the importance of extra-organizational communities of practice which may involve geographically co-located and distributed members. Here individuals come together to create a film and once this is achieved they disperse, yet they remain members of a film making community even when they are no longer employed by a film producing business organization (see, for instance, DeFillippi & Arthur, 1998). The shared enterprise, mutual engagement, and shared repertoire of the film making community are brought to bear in the temporary project of the production of a film, but it is in these extra-organizational communities of practice that new members gain legitimate peripheral participation and over time become full participants. Employment opportunities will come and go but membership of a community of practice becomes an important constant in the lives of certain workers in the current accelerated business environment (Roberts, 2006).

Communities of practice would then appear to hold value in terms of understanding learning and knowledge generation beyond the boundaries of commercial organizations at various spatial scales. However, it is questionable as to whether what are conceptualized as communities of practice in many empirical studies are communities of practice in the sense of Lave and Wenger's (1991) original elaboration. The original conceptualization was very much one of learning and knowledge creation embedded and situated in social practices. But the notion of communities of practice has been stretched and has become so flexible that it is now applied to a wide variety of learning and knowledge creating activities. As Duguid (2008) notes, the communities of practice approach is a theory under continual social construction.

Many of the activities referred to as communities of practice by management practitioners and academics do not include situated social practice, but rather dislocated practice, with members being separated in time and/or space. Also, although much of the interaction incorporated in studies of communities of practice may be situated, in the sense that it occurs face-to-face, it may not be directly linked to the process of learning and knowledge

90 Joanne Roberts

Table 4.4 Varieties of Knowing in Action

Activity	Type of knowledge	Social Interaction				Innovation	Organizational dynamic
		Proximity/ nature of communication	Temporal aspects	Nature of social ties		
Craft/ task based	Aesthetic, kinaesthetic, and embodied knowledge	Knowledge transfer requires co-location—face-to-face communication, importance of demonstration	Long-lived and apprenticeship-based Developing socio-cultural institutional structures	Interpersonal trust—mutuality through the performance of shared tasks	Customized, incremental	Hierarchically managed Open to new members
Professional	Specialized expert knowledge acquired through prolonged periods of education and training Declarative knowledge Mind-matter and technologically embodied (aesthetic and kin-aesthetic dimensions)	Co-location required in the development of professional status for communication through demonstration. Not as important thereafter	Long-lived and slow to change. Developing formal regulatory institutions	Institutional trust based on professional standards of conduct	Incremental or radical but strongly bound by institutional/ professional rule Radical innovation stimulated by contact with other communities	Large hierarchically managed organizations or small peer managed organizations Institutional restrictions on the entry of new members

Communities as Spaces of Innovation

Activity	Type of knowledge	Social Interaction — Proximity/ nature of communications	Temporal aspects	Nature of social ties	Innovation	Organizational dynamic
Epistemic/ creative	Specialized and expert knowledge, including standards and codes, (including meta-codes) Exist to extend knowledge base Temporary creative coalitions; knowledge changing rapidly	Spatial and/or relational proximity. Communication facilitated through a combination of face-to-face and distanciated contact	Short-lived drawing on institutional resources from a variety of epistemic/ creative fields	Trust based on reputation and expertise, weak social ties	High energy, radical innovation	Group/project managed Open to those with a reputation in the field Management through intermediaries and boundary objects
Virtual	Codified and tacit from codified Exploratory and exploitative	Social interaction mediated through technology— face-to-screen Distanciated communication Rich web-based anthropology	Long and short lived Developing through fast and asynchronous interaction	Weak social ties; reputational trust; object orientation	Incremental and radical	Carefully managed by community moderators or technological sequences Open, but self regulating

Source: Amin and Roberts (2008b, 357).

creation in practice. Social interactions in professional associations, for instance, may affect the knowledge or learning of individual members, but that learning does not necessarily occur through the socially situated practice of a number of members working together. So there is a sense in which social interaction can occur without a direct impact on practice. This type of social interaction is dislocated from practice.

Just as Lindkvist (2005) develops an alternative, though complementary, view of communities of practice in the form of collectivities of practice, to refer to temporary groups or project teams concerned with knowledge creation and exchange, there is a need to differentiate between communities of practice of different sizes and spatial scales and the nature of practice. There are then different types of knowledge creating and transferring communities. Based on an extensive review of the communities of practice literature, Amin and Roberts (2008b, 357) present a typology to differentiate between various sorts of knowing in action (Table 4.4). They identify four types of knowing associated with four separate, though often overlapping, activities: Craft/task-based, Professional, Epistemic/creative, and Virtual. As Amin and Roberts (2008b) note, the development of knowledge in communities forming around these four activities requires distinctive patterns of social interaction and spatial characteristics and they give rise to different sorts of innovation. However, what is clear is that activities taking place in a variety of spatial distinctive communities do give rise to innovative activity.

Hence, it is possible to employ the notion of communities of practice to understand the social dynamics of innovation beyond locationally bound knowledge generating activity. Moreover, in professional, epistemic, and virtual communities relational proximity can facilitate high levels of interaction across distance allowing the community to become a place for innovative activity that is dislocated from geographical/territorial space. In this sense then, communities can add to understandings of the spatial dimensions of innovation both by revealing the micro dimensions of activities in terms of the social dynamics in specific locations and by highlighting the significance of relational proximity. By bridging across distance, such communities link spatially distributed innovative capacities, thereby allowing greater opportunities for the development of new knowledge. The mediated sociality occurring in virtual communities can be no less powerful than the face-to-face sociality of co-located communities.

REFLECTIONS ON COMMUNITIES AS SPACES OF INNOVATION

While the community of practice literature is increasingly popular (Wenger et al., 2002; Saint-Onge & Wallace, 2003), and it has become influential beyond the business and management field, it has attracted criticism. For instance, the marginalization of the issue of power has been noted (Contu

& Willmott, 2003; Fox 2000), as has the failure to take into account preexisting conditions such as habitus and social codes (Mutch, 2003). In addition, Handley et al. (2006) highlight the way that the communities of practice approach has moved from its original focus on situated learning. The use of the term 'community,' which embodies positive connotations and is open to multiple interpretations, is also viewed as problematic (Lindkvist, 2005; Roberts, 2006). Some of these general criticisms have implications for the extent to which communities may be effective spaces of innovation. For example, power relations within communities may prevent the adoption of new practices where they threaten the status of core members as studies of professional communities of practice in the healthcare sector demonstrate (Swan et al., 2002).

It is also important to note that the bonding social capital, which underpins reciprocity, loyalty, and solidarity within communities, can encourage an inward looking stance and resistance to change. This can have negative consequences for the innovative potential of communities. Social bonding capital may, however, be balanced by social bridging capital which supports an open and outward looking stance providing scope to engage with other communities in a productive and reciprocal manner. As Putman (2000, 23) notes '[b]onding social capital constitutes a kind of sociological superglue, whereas bridging social capital provides a sociological WD-40.' In a sense, bridging social capital underpins the strength of weak ties (Granovetter, 1973) and allows innovative community members to access knowledge from a diverse range of other communities. Moreover, it is often at the intersections between communities that innovation occurs.

Nevertheless, the degree to which communities are open to new ideas and the changes that these may bring about vary. For instance, a community of practice in a professional sector may be open to new knowledge from other communities of practice in the profession, but closed to knowledge from other professions or non professional sources. Openness to outside perspectives will be mediated by a range of differences, including norms, values, cultures, ethics, and so on. For knowledge to be successfully mobilized, transferred, or shared, a degree of alignment between insiders and outsiders is crucial. Inter and extra organizational communities of practice can be sources of such alignment and the benefits it offers.

For instance, Wenger et al. (2002, 220) argue that 'interorganizational communities of practice help to maintain internal expertise while strengthening relationships with outsourcing partners' (p. 220). Hence communities of practice that exist independently of business organizations may take on an increasingly important role in the creation and transfer of knowledge. Furthermore, Kimble and Hildreth (2004) argue that workers increasingly operate in an individualistic world of weak ties where resources are frequently obtained through personal networks and individual relationships rather than through organization based communities. Individuals belong to a variety of communities of practice, some internal to their work organization, while others will be external, arising from their personal and professional networks. For business organizations then there are both challenges and opportunities arising from the participation of employees in external communities. On the one hand, they offer a source of additional knowledge, but on the other they raise the risk of knowledge leaking to competitors. Nevertheless, in fast changing markets there is an increasing trend toward open innovation (Chesbrough 2003) as well as a recognition that innovation does not always occur within the firm (Gabriel 2005). For instance, lead user communities have been identified as important contributors to product development in a variety of sectors (Von Hippel, 2005), and companies are increasingly seeking to harness their input through, for instance, specific social software (Burger-Helmchen & Cohendet, 2011).

The value of communities of practice in the promotion of learning and knowledge generation will vary according to the broad socio-cultural context (Roberts, 2006). For instance, national competitiveness deriving from knowledge creating and sharing capabilities may vary depending on nation-specific socio-cultural characteristics, such as, levels of trust or the relative position of the individual versus the community. The community of practice as a tool of learning and knowledge generation may well be more successful in those regions and nations that have a strong community spirit, so long as they remain open to outside influences, compared to those nations that have a weak community spirit. For instance, in relation to Hofstede's (1991) study of national culture, one might expect that a nation characterized by collectivism might find that the community of practice is a more effective knowledge creation and dissemination strategy than in nations characterized by individualism. If this is the case, then a country wishing to excel in the knowledge-based era through the development of communities of practice will need to promote community in the wider society through education and social infrastructures. Consequently, the broad national system of innovation will influence the success of communities of practice as a mechanism for knowledge transfer and generation. It is also important to recognize that a nation's innovative capacities may also be influenced by the presence of immigrant communities that are linked to a widely dispersed diaspora or from returnee entrepreneurs like those highlighted by Saxenian (2006) in the high tech sector. These groups bring with them fresh socio-cultural perspectives that have the potential to influence a regional or national innovation system. In the highly globalized world of the 21st century, those involved in innovation have a wide variety of socio-cultural knowledge from which to draw.

Yet, as Nooteboom (2008) notes, for creative interaction between communities there is a need for sufficient cognitive proximity to allow productive communication. Such cognitive proximity may imply a degree of shared social capital. Too little cognitive proximity will undermine the ability of members from different communities to engage with one another—there will be insufficient common knowledge from which to develop mutual understandings. However, where cognitive proximity is too high there will be little to be gained from interaction since the knowledge sets of both parties will be too similar to offer opportunities for creativity arising from cognitive dissonance. Hence, a balance between cognitive distance and cognitive proximity is required to allow productive communication between the communities yet simultaneously provide sufficient cognitive dissonance to spark creativity. So while communities may offer a means of facilitating innovation, inter community innovation may be limited by the extent to which the interaction is characterized by cognitive dissonance rather than cognitive resonance. Nooteboom's (2008) analysis underlined the need for bridging social capital to link different communities and the significance of weak ties for the creative potential of inter-community activity.

CONCLUSION

This chapter began by considering the turn to community before exploring early conceptualizations of communities of practice (Lave and Wenger, 1991; Wenger, 1998). The approach has been applied to contexts far beyond its original setting of situated learning in apprenticeships. Hence, it is important to differentiate between various types of communities. Amin and Roberts's (2008b) typology of knowing in action is presented (Table 4.4) to illustrate the differences between spatial reach and the social dynamics evident in particular sorts of communities. Such differences impact on the innovative capacities of communities.

Nevertheless, the community of practice does offer a mechanism for the exploration of knowledge transfer and generation in various spatial contexts. It offers insight into the micro aspects of interaction between the individual parties involved; as such, it contributes to understandings of the social dynamics of innovation. In addition, the communities of practice approach has the potential to complement other frameworks, such as regional and national systems of innovation, industrial districts, clusters, innovative milieus, and networks, which seek to explain knowledge transfer and generation at regional and national levels. By revealing how the social situated interaction within such environments contributes to knowledge transfer and generation, communities of practice add another layer to our understanding of innovation and learning processes. Indeed, the incorporation of spatially distributed communities, which engage in distanciated interaction and generate bridges across which distributed knowledge can flow into local innovation activity, deepens our understanding of innovation in a globalized world (Gertler & Levitte, 2005; Boschma, 2005; Bathelt, Malmberg, & Maskell, 2004; inter alia). In this way, local actors can tap into a diverse range of socio-cultural resources to assist in the generation of new knowledge.

Analyses of organizational specific communities of practice suggest that managers can do much to promote their successful development (Brown and Duguid, 2001a; Wenger, et al., 2002; Saint-Onge & Wallace, 2003). Wenger et al. (2002) claim that communities of practice must be cultivated if business organizations are to fully exploit the opportunities they offer. However, such cultivation requires resources that are not readily available to small and medium-sized organizations. As Roberts (2006) notes, while very small organizations and professional practices may be amenable to the spontaneous development of communities of practice, small and medium sized enterprises (SMEs) with limited resources may be less able to mobilize communities as a means of knowledge transfer and innovation.

Policy-makers, while not being able to create local, regional, or national communities of practice, can contribute to their construction through the provision of supportive institutions and infrastructures. For example, public support for professional, trade, and interest-based associations in the form of tax relief on membership fees and grants for the knowledge generating activities of such groups would encourage their development and sustainability. Moreover, public funding together with commercial sponsorship to support the creation of such associations where a need is identified could prove to be valuable long-term investments. In addition, support for SMEs to promote their ability to cultivate communities of practice would increase their capacity to mobilize resources for knowledge transfer and generation. An example of an existing U.K. government initiative that supports the development of communities forming between universities and industry is the Knowledge Transfer Partnership programme. Gertner et al. (2011) studied the formation of communities of practice in three such partnerships and identified their positive knowledge transfer and innovation outcomes.

Even though communities of practice of various sorts have been subject to a significant amount of research (Amin & Roberts, 2008b; Murillo, 2011), considerations of the spatial dimensions of the innovation they facilitate remains underdeveloped and in need of further investigation. In particular, the capacity of communities of practice to reach across various spatial scales requires further attention. Developing knowledge of this aspect of communities of practice would also provide insights for managers and policy-makers to enhance their ability to promote the innovative capacity of community.

NOTES

1. HTTP://en.wikipedia.org/wiki/Wikipedia. Accessed February 12, 2012.

REFERENCES

Adler, P.S. (2001) Market, hierarchy, and trust: The knowledge economy and the future of capitalism. *Organization Science* 12 (2), 215–234.
Amin, A. (2002) Spatialities of globalisation. *Environment and Planning A*, 34, 385–399.

Amin, A., & Cohendet, P. (2004) *Architectures of knowledge: Firms, capabilities, and communities*. Oxford: Oxford University Press.
Amin, A., & Roberts, J. (2008a) The resurgence of community in economic thought and Practice. In Amin, A., & Roberts, J. (eds.), *Community, Economic Creativity and Organization*. Oxford: Oxford University Press, pp. 11–34.
Amin, A., & Roberts, J. (2008b) Knowing in action: Beyond communities of practice. *Research Policy*, 37 (2), 353–369.
Barley, S., & Orr, J. (eds.). (1997) *Between craft and science: Technical work in the United States*. Ithaca: Cornell University Press.
Bathelt, H., Malmberg, A., &Maskell, P. (2004) Clusters and knowledge: Local buzz, global pipelines and the process of knowledge creation. *Progress in Human Geography* 28 (1), 54–79.
Becattini, G. (1990) The Marshallian industrial district as a socio-economic notion. In Pyke, F. et.al. (eds.), Industrial districts and inter-firm cooperation in Italy. Geneva: International institute for labour studies, pp. 37–51.
Becattini, G. (2004) *Industrial districts*. Cheltenham: Edward Elgar. .
Benner, C. (2003) Learning communities in a learning region: The soft infrastructure of cross-firm learning in networks in Silicon Valley. *Environment and Planning A*, 35, 1809–1830.
Bjorg, A., and Isksen, A. (1997) Location, agglomeration and innovation: Towards regional innovation systems, in Norway. *European Planning Studies*, 5 (3), 299–331.
Blackler, F. (1995) Knowledge, knowledge work and organizations: An overview and interpretation. *Organization Studies*, 16 (6), 1021–1046.
Boland, R.J., & Tenkasi, R.V. (1995) Perspective making and perspective taking in communities of knowing. *Organization Science*, 6 (4), 350–372.
Borzillo, S., Aznar, S., & Schmitt, A. (2011) A journey through communities of practice: How and why members move from the periphery to the core. *European Management Journal*, 29, 25–42.
Boschma, R.A. (2005) Proximity and innovation: A critical assessment. *Regional Studies* 39 (1), 61–74.
Bowles, S., & Gintis, H. (2002) Social capital and community governance. *The Economic Journal*, 112, F419–F436.
Bresnahon, T.F., & Gambardella, A. (2004) *Building high-tech clusters: Silicon Valley and beyond*. Cambridge: Cambridge University Press.
Brown, J.S., & Duguid, P. (1991) Organizational learning and communities of practice: Towards a unified view of working, learning and innovation. *Organization Science*, 2, 40–57.
Brown, J.S., & Duguid, P. (1998) Organizing knowledge. *California Management Review*, 40 (3), 90–111.
Brown, J.S., & and Duguid, P. (2001a) Structure and spontaneity: Knowledge and organization. In Nonaka, I., & Teece, D. (eds.), *Managing Industrial Knowledge*. London: Sage Publication, pp. 44–67.
Brown, J S., & and Duguid, P. (2001b) Knowledge and organization: A social-practice perspective. *Organization Science*, 12 (2), 198–213.
Burger-Helmchen, T., & Cohendet, P. (2011) 'User communities and social software in the video game industry. *Long Range Planning*, 44 (5–6), 317–343.
Camagni, R.P. (1995) The concept of innovative milieu and its relevance for public policies in European lagging regions. *Papers in Regional Science*, 74 (4), 317–340.
Carlile, P.R. (2002) A pragmatic view of knowledge and boundaries: Boundary objects in new product development. *Organization Science*, 13 (4), 442–455.
Chesbrough, H.W. (2003) *Open innovation: The new imperative for creativity and profiting from technology*, Boston: Harvard Business School Press.

Coe, N.M., & Bunnell, T.G. (2003) "Spatializing" knowledge communities: Towards a conceptualisation of transnational innovation networks. *Global Networks*, 3 (4), 437–356.

Cohendet, P., Grandadam, D., & Simon, L. (2010) The anatomy of the creative city. *Industry and Innovation*, 17 (1), 91–11.

Cohendet, P., & Simon, L. (2008) Knowledge-intensive firms, communities and creative cities. In Amin, A., & Roberts, J. (eds,), *Community, Economic Creativity, and Organization*. Oxford: Oxford University Press, pp. 227253.

Contu, A., & Willmott, H. (2003) Re-embedding situatedness: The importance of power relations in learning theory. *Organization Science*, 14 (3), 283–296.

Conway, C., Dawley, S., & Charles, D. (2005) *An investigation of the learning dynamics involved within the creative industries sector in the Newcastle City Region*. Working Paper, June, Centre for Urban and Regional Development Studies, Newcastle University.

DeFillippi, R. (2002) Organizational models for collaboration in the new economy. *Human resources planning*, 25 (4) 7–18.

DeFillippi, R.J., & Arthur, M.B. (1998), Paradox in project-based enterprise: The case of film making. *California Management Review*, 40 (2), 125–139.

Duguid, P. (2008) Prologue, community of practice then and now. In Amin, A., & Roberts, J. (eds.), *Community, Economic Creativity and Organization*. Oxford: Oxford University Press, pp.1–10.

Etzioni, A. (1996) *The new golden rule: Community and morality in a democratic Society*. New York: Basic Books.

Etzioni, A. (2000) Creating Good Communities and Good Societies, *Contemporary sociology*, 29 (1), 188–195.

Faulconbridge, J.R. (2010) Global architects: learning and innovation through communities and constellations of practice. *Environment and Planning A*, 42 (12), 2842–2858.

Foss, N.J. (2006) *Strategy, economic organization, and the knowledge economy: The coordination of firms and resources*. Oxford: Oxford University Press.

Fox, S. (2000) Communities of practice, Foucault and Actor-Network Theory. *Journal of Management Studies*, 37 (6), 853–867.

Gabriel, R.P. (2005) *Innovation happens elsewhere: Open source as business strategy*. San Francisco: Morgan Kaufmann,.

Gertler, M.S. (2008) Buzz without being there? Communities of practice in context. In Amin, A., & Roberts, J. (eds.), *Community, economic creativity and organization*. Oxford: Oxford University Press, pp.203–226.

Gertler, M.S., & Levitte, Y.M. (2005) Local nodes in global networks: The geography of knowledge flows in biotechnology innovation. *Industry and Innovation*, 12 (4), 487–507.

Gertner, D., Roberts, J., & Charles, D. (2011) University-industry collaboration: A CoPs approach to KTPs. *Journal of Knowledge Management*, 15 (4), 625–647.

Gherardi, S., Nicolini, D., & Odella, F. (1998) Towards a social understanding of how people learn in organizations. *Management Learning*, 29 (3), 273–298.

Granovetter, M.S. (1973) The strength of weak ties. *American Journal of Sociology*, 78 (6), 1360–1380.

Haas, P. (1992). Introduction: Epistemic communities and international policy coordination. *International Organization*, 46 (1), 1–37.

Håkansson, Hakan (ed.) (1987) *Industrial technological development—A network approach*. London: Croom Helm.

Hall, H., & Graham, D. (2004) Creation and recreation: Motivating collaboration to generate knowledge capital in online communities. *International Journal of Information Management*, 24 (3), 235–246.

Handley, K., Sturdy, A., Fincham, R., & Clark, T. (2006) Within and beyond communities of practice: Making sense of learning through participation, identity and practice. *Journal of Management Studies*, 43 (3), 623–639.
Handy, C. (1998) *The age of unreason*. Cambridge: Harvard Business School Press.
Hara, N., Shachaf, P., & Stoerger, S. (2009) Online communities of practice typology revisited. *Journal of Information Science*, 2009, 35, 740–757.
Hofstede, G. (1991) *Cultures and organizations: Software of the mind*. London: McGraw-Hill.
Howells, J. (1999) Regional innovation systems? In Archibugi, D., Howells, J., & Michie, J. (eds.), *Innovation systems in a global economy*. Cambridge: Cambridge University Press, pp. 67–93.
Johnson, C.M. (2001) A survey of current research on online communities of practice. *Internet and Higher Education*, 4, 45–60.
Kimble, Chris, & Hildreth, P. (2004) Communities of practice: Going one step too far? *Proceedings 9e L'AIM*, Evry, France, May.
Knorr-Cetina, K. (1999) *Epistemic cultures: How the sciences make sense*. Chicago: University of Chicago Press.
Lave, J., & Wenger, E. (1991) *Situated learning: Legitimate peripheral participation*. Cambridge: Cambridge University Press.
Leadbeater, C. (2008) *We-Think: Mass innovation, not mass production*. London: Profile Books Ltd. .
Lesser, E.L., & Storck, J. (2001) Communities of practice and organizational performance. *IBM Systems Journa*, 40 (4), 831–841.
Lindkvist, L. (2005) Knowledge communities and knowledge collectivities: A typology of knowledge work in groups. *Journal of Management Studies*, 42 (6), 1189–1210.
Lundvall, Bengt-Åke. (1992) Introduction. In Bengt-Åke, Lundvall (ed.), *National systems of innovation: Towards a theory of innovation and interactive learning*. London: Pinter, pp. 1–19.
Malmberg, A., & Maskell, P. (2002) The elusive concept of localization economies: Towards a knowledge-based theory of spatial clustering. *Environment and Planning A*, 34 (3), 429–449.
Markusen, A. (1996) Sticky places in slippery space: A typology of industrial districts. *Economic Geography*, 72 (3), 293–313.
Miles, R.E., Snow, C.C., Mathews, J.A., Miles, G., & Coleman, H.J. (1997) Organizing in the knowledge age: Anticipating the cellular form. *Academy of Management Executive*, 11 (4), 7–20.
Mintzberg, H. (2009) Rebuilding companies as communities. *Harvard Business Review*, July–August, 140–143.
Murillo, E. (2011) Communities of practice in the business and organization studies literature. *Information Research*, 16 (1).http://informationr.net/ir/16-1/paper464.html. Accessed February 12, 2012.
Mutch, A. (2003) Communities of practice and habitus: A critique. *Organization Studies*, 24 (3), 383–401.
Nelson, R.R. (1993) A retrospective. In Nelson, R.R. (ed.), *National innovation systems: A comparative analysis*. New York: Oxford University Press, pp. 505–523.
Nooteboom, B. (2008) Cognitive distance in and between communities of practice and firms: Where do exploitation and exploration take place, and how are they connected? In Amin, A., & and Roberts, J. (eds.), *Community, economic creativity and organization*. Oxford: Oxford University Press, pp. 123–147.
Orr, J.E. (1996) *Talking about machines: An ethnography of a modern job*. Ithaca, London: IRL Press, an imprint of Cornell University Press.

O'Sullivan, D. (2009) *Wikipedia: A new community of practice?* Farnham: Ashgate.

Pan, S.L., & Leidner, D.E. (2003) Bridging communities of practice with information technology in pursuit of global knowledge sharing. *Journal of Strategic Information Systems*, 12, 71–88.

Porter, M.E. (1998) Clusters and the new economics competition. *Harvard Business Review*, November–December, 77–90.

Putnam, R.D. (2000) *Bowling alone: The collapse and revival of American community*. New York: Simon & Schuster Paperbacks.

Quinn, B.J. (1992) *Intelligent enterprise: A knowledge and service based paradigm for industry*. New York: Free Press.

Roberts, J. (2006) Limits to communities of practice. *Journal of Management Studies*, 43 (3), 623–639.

Roberts, J. (2010a) Community and international business futures: Insights from software production. *Futures*, 42 (9), 926–936.

Roberts, J. (2010b) Communities of management knowledge diffusion. *Prometheus: Critical Studies in Innovation*, 28 (2), 111–132.

Roberts, J., & Armitage, J. (2006) From organization to hypermodern organization: On the accelerated appearance and disappearance of Enron. *Journal of Organizational Change Management*, 19 (5), 558577.

Saint-Onge, H., & Wallace, D. (2003) *Leveraging communities of practice for strategic advantage*. London, New York: Butterworth Heinemann.

Saxenian, A.L. (2006) *The new Argonauts: Regional advantage in a global economy*. Cambridge: Harvard University Press.

Sole, D., & Edmondson, A. (2002) Situated knowledge and learning in dispersed teams. *British Journal of Management*, 13, S17–S34.

Star, S.L., & Griesemer, J.R. (1989) Institutional ecology, "Translations" and boundary objects: Amateurs and professionals in Berkeley's Museum of Vertebrate Zoology, 1907–39. *Social Studies of Science*, 19 (3), 387–420.

Storper, M. (2008) Community and economics. In Amin, A., & Roberts, J. (eds.), *Community, economic creativity and organization*. Oxford: Oxford University Press, pp. 37–68.

Swan, J., Scarbrough, H., & Robertson, M. (2002) The construction of "Communities of Practice" in the management of innovation'. *Management Learning*, 33 (4), 477–496.

Teigland, R. (2000) Communities of practice in a high-technology firm. In Birkinshaw, J., & Hagström, P. (eds.), *The flexible firm: Capability management in network organizations*. Oxford: Oxford University Press, pp. 126–146.

Tonnies, F. (1963) *Community and association*. Trans. Charles P. Loomis. New York: Harper and Row.

Von Hippel, E. (2005) *Democratizing innovation*. Cambridge: MIT Press.

Wenger, E. (1998) *Communities of practice: Learning, meaning, and identity*. Cambridge: Cambridge University Press.

Wenger, E.C. (2000) Communities of practice and social learning systems. *Organization*, 7 (2), 225–246.

Wenger, E.C., & and Snyder, W.M. (2000) Communities of practice: The organizational frontier. *Harvard Business Review*, January–February, 139–145.

Wenger, E. C. and Snyder, W. M. (2000) 'Communities of Practice: The Organizational Frontier'. *Harvard Business Review*, January-February, 139–145.

Wenger, E.C., McDermott, R., & and Snyder, W.M. (2002) *Cultivating communities of practice: A guide to managing knowledge*. Boston: Harvard Business School Press.

Part II
Innovation and Social Capital
A Reconsideration of Conceptual and Methodological Dilemmas

5 National and Regional Innovation Capacity Through the Lens of Social Capital
A Qualitative Meta-Analysis of Recent Empirical Studies

Frane Adam

INTRODUCTION

The purpose of this chapter is to provide a systematic overview of existing studies of social capital within the framework of national and regional innovation systems and to attempt to ascertain whether there have been any related new insights or methodological shifts in the last decade[1]. The reconstruction of empirical findings reveals quite an ambiguous picture. It seems that only a radical modification of research designs and methods can give a fresh impetus and allow us to gain a deeper understanding of social capital's role and/or its dimensions in (regional) innovation processes. Moreover, the qualitative meta-analysis of recent studies was conducted in order to discern possible and more or less reliable and theoretically interesting common denominators regarding the interrelationship of social capital and innovation.

Innovative performance, mainly measured in terms of R&D intensity (as an input variable) and patenting (as an output variable) at different spatial levels and sectors, raises interesting and theoretically challenging issues. Also evident is the policy relevance of the search for national and regional competitive advantages—on the basis of their spatially determined innovation systems which in many cases are firmly embedded in global contexts. Some nations and regions are known to have been able to exploit their geographical proximity and concentration of communication flows, and to create collaborative structures, knowledge sharing, and interactive learning, resulting in intensive innovation activity. Multi-disciplinary research has shown growing interest in the socio-cultural and cognitive factors that may accelerate innovation activity, knowledge transfer, and the formation of a creative environment. In the last decade or so, one of the emphases in this context has been on investigating social capital's role in innovation-capacity building, at both national and regional levels as well as at the level of companies and micro networks (for an overview, see Zheng, 2008). We will chiefly limit ourselves to recent empirical studies conducted mainly within Europe, bearing in mind that significantly fewer studies dealing explicitly with social capital have been conducted (or published) than would be expected, especially when we consider the overall trend of its rapid reception and applications (Woolcock, 2010; Yin & Chiang, 2010). However, many more studies can be found that employ concepts usually connected with social capital, such as social networks, clusters, trust, or collaborative structures, albeit without (explicit) reference to the concept of social capital. It seems that some authors may prefer to avoid this concept. In contrast, we also encounter studies that use this notion (too) frequently and yet employ indicators which may or may not truly measure the intended dimensions or aspects of social capital. This uncertainty seems to be linked to the inherent ambiguity and limitation of this concept.

A CONTESTED CONCEPT

Many scholars recognise that in the last 20 years (especially in the last decade) the concept of social capital has enjoyed extraordinary reception and application in various fields of social research and beyond (in the media, politics, etc.). Originally a marginal concept, it has become a mainstream notion and a "minor industry in social science," despite its opaqueness and catch-all character (Häuberer, 2011; Van Deth, 2003; Haynes, 2009). It seems that this great attractiveness and broad application across different disciplines (see the bibliometric analysis by Yin & Chiang, 2010) have come at a price: the concept has been watered down and excessively quantified, and relatively uncoordinated empirical research has been carried out. In a recent retrospective analysis of the 'genealogy' and emerging epistemic status of social capital, two features are stressed: routinisation and controversial reception, summarised in the syntagm of "an essentially contested concept" (Woolcock, 2010). Routinisation refers to its incorporation in social science vocabulary, in handbooks and college courses, journal submissions—and we can add its inclusion in empirical research mostly on the basis of 'positivistic' reasoning. This can be defined as a 'spontaneous' ideology (doctrine) of certain groups of scientists that takes for granted the production of quantifiable data as a sufficient and necessary basis for statistical processing and the creation of new knowledge.

The second feature refers to two elements: first, to the lack of a consensus on an operational definition and, second, to the ability to facilitate dialogue between disciplines. Put differently, the utility of the social-capital approach "rests less on its capacity to forge an inherently elusive scholarly or policy consensus on complex issues than on its capacity to facilitate constructive dialogue about agreements and disagreements between groups who would otherwise rarely (if ever) interact" (Woolcock, 2010, 469). While it may be accepted that these two features are interconnected, it is quite clear they generate different consequences. It can even be argued that routinisation (which may correspond to the notion of "normal science" in Kuhn's sense) renders the presumed dialogue across disciplinary boundaries difficult. On the other hand, the author (Woolcock) of this diagnosis of the state of the art of social-capital application unfortunately does not provide any concrete evidence of the content and results of the alleged dynamic dialogue and reflection. It may be said that his analysis lacks insight into research designs and empirical findings since it largely focuses on an overview of its reception by various authors and of its adoption in different disciplines (especially political science).

Why is social capital "an essentially contested concept"? There are many reasons. The concept has been widely accepted and adopted by disciplines with different traditions, methodological standards, and thematic orientations. This has contributed to the continuing ambiguity and quite diversified understandings and applications of this concept. One element of its contestation and controversial reception involves its theoretical status. It can be argued that social capital is less a theoretical approach than primarily a *framing and heuristic concept* (for more on the role of framing in research design, see Ragin & Amoroso, 2011).[2] Expressed differently: in *sensu strictu,* it is very difficult to talk about a social-capital theory since it does not have a clear "object".[3] The basic operational definition points to other concepts such as trust, social network, civic participation (civil society), or collective action. These concepts had all been theoretically elaborated before the 'invention' and transposition of social capital to them. Some authors wonder whether this is a case of complementarity or redundancy (e.g., Fromhold-Eisebith, 2004). Yet the issue has to do with the contested and contradictory reception of social capital, as reflected in empirical research including research into innovation. It can be argued that unresolved problems and ambiguity concerning the level of definition and operationalisation tend to lead to methodological shortcomings and non-comparable research designs.

SOCIAL CAPITAL: A REDUNDANT OR A COMPLEMENTARY CONCEPT?

According to the prevailing definitions of social capital, its externalities resulting from trust, density of informal (network) contacts, or (active) membership in voluntary associations may generate a kind of *capital analogous to economic or human capital.* In this regard, social capital has its roots in Bourdieu's theory of symbolic capitals—which forms part of a broader theory of stratification and class reproduction—and partially in Becker-Colleman's extension of human-capital theory to social capital.[4] The social-capital approach provides a new frame of reference; it is a mobilising and integrating concept in the sense that the use of social-capital terminology may also imply the intention to operate with a broader set of its dimensions/antecedents. Here the comparison of two similar concepts is very instructive:

'innovative milieu', which is older and theoretically elaborated in the work of GREMI,[54] and 'social capital' in the framework of the innovation process. Both point out the advantage of a dense network of social contacts and trustful relationships for innovation performance. According to Fromhold-Eisebith (2004), the difference should lie in the character of the networks. The milieu approach is based on "trustful relationships between heterogeneous actors," while the social-capital approach is supposedly related to "a foundation of homogeneity, also promoted by the often closed and institutionalized nature of relevant networks." The author concludes that the two concepts are complementary and not redundant (Fromhold-Eisebith, 2004, 754). But the fact remains that the milieu approach was invented in order to describe the structure of social relations and environmental characteristics conducive to innovation activities, while social capital is a concept that emerged later and has only recently been used as a factor to explain the dynamics of knowledge sharing and innovativeness.

However, many (most?) supporters of network theory as well as the social-capital approach would not agree with the homogeneous and bonding nature of social capital.[6] Instead, in the eyes of prominent researchers the most productive form of social capital refers to heterogeneous networks and terms like 'weak ties' 'bridging social capital, and 'structural holes.' Or, to be more precise with regard to innovation: in the early stages of an innovation process (invention), it is more important to be exposed to a flow of diverse ideas, while in later stages it is advantageous to have support within the team or firm; and the same is true with regard to the radicalness of innovations (for more on different types and levels of inventions/innovations, see Cooke, 2008; Landry, Amara, & Lamari, 2002). In the case of the above comparison, we clearly have to do merely with the distinction between bonding and bridging social capital. It is obvious that both concepts (milieu and social capital) overlap greatly and are redundant (also see criticism expressed in a paper by Tura & Harmaakorpi, 2003, of Fromhold-Eisebith's ideas of complementarity and the equation of social capital with stability).

Another author writes that "innovative outcomes may be more the result of collaborative structure rather than that of social capital" (Zheng, 2008, 179). Are collaborative structures not an intentional product or by-product of social capital?[7] What then is social capital and its function if it is not conceived as a catalyst for cooperative forms and flexible organisational structures? Woolcock provides the most condensed formula for social capital: "Social capital is a proto-theory of human co-operation" (Woolcock, 2010). This statement is very important since it can be interpreted on one hand as recognition that social capital is not a strictly theoretical concept (it has a proto-theoretical meaning) and, on the other, as emphasising its role and function as a provider or facilitator of cooperative forms of action. If social capital is still in its proto-theoretical form after 20 years, then it is probably not very productive to further seek its genuine theoretical foundation and integration. It seems better to be satisfied with the less ambitious, proto-theoretical and framing concept, and to strive for more coordinated and deliberative empirical research in this specific field.[8]

Some definitions of social capital, particularly those based on the network approach, emphasise that the mere placement of an individual or group in an interactive system or social network provides certain benefits or assets. Even if we accept as a basis the thesis of social capital as a resource (asset) that enhances cooperative relations and, as a result, enables the achievement of certain individual and collective goals, it is clear that the forms of cooperation and goals are highly diverse, varying from the everyday (mundane) to those more complex and specialised. In other words, a distinction must be made between primordial social capital, which is grounded in the basic structure of everyday life (or the so-called life-world), and more complex and specialised forms of social capital or 'higher' forms of cooperation and more complex collaborative structures (among these we can also cite to some extent Putnam's approach, which emphasises civic participation as the main form of specialised social capital).

Similar to the phenomenological analysis of knowledge distribution, with its distinction between simple and complex knowledge distribution, is the distinction between the simple distribution of social capital and complex distribution of (specialised) social capital. However, primordial social capital or its simple distribution includes variations which are the result of: (1) different positions (and 'investments') of the individual within the egocentric network or the type of bonds; and (2) the integration of these networks within a system of stratification and social power. In the context of innovation performance, complex and specialised social capital is needed in order to increase the motivation and ability for flexible organisational forms such as teamwork (Costa, Bijlsma-Frankema, & de Jong, 2009), interactive learning, project and networking management, interdisciplinary collaboration, or intermediary institutions.

Within regional innovation systems, functional connections and cooperative structures composed of companies ('industrial districts') or of companies and producers of knowledge (universities, institutes) and intermediary institutions are seen as the basis of propulsive development (for further reading, see Asheim, Coenen, & Svensson-Henning, 2003; Cooke, 2008; Isaksen, 2009). In this framework, more complex and specialised forms of cooperation and coordination appear, enabled and promoted by social capital, although of course there are other conditions and prerequisites for their functioning. The problem is that the majority of research examining the impact of social capital on regional innovation performance is based on primordial forms of social capital (like informal contacts) or civic participation, while not enough attention is given to other, specialised, concrete collaborative patterns. In the set of 18 studies (see the Appendix 5.1), only three focus on such types (Fromhold-Eisebith, 2004; Kallio, Harmaakorpi, & Pihkala, 2010; Bruno et al., 2008).

CONVERGENT AND DIVERGENT FINDINGS REGARDING SOCIAL CAPITAL'S IMPACT ON NATIONAL REGIONAL INNOVATION CAPACITY

It became clear during the collection of empirical studies dealing with the determinants (facilitators) of innovation performance and knowledge transfer that some authors use the term social capital and others not, even though they address the same factors, i.e., social networks, structural holes, weak and strong ties (Fleming, Mingo, & Chen, 2007; Liebowitz, 2007), "cooperation networks" (Isaksen, 2009; Russo, & Rossi, 2008; Ahrweiler, ed., 2010), different types of proximity (Boschma, 2005), and trust (Nijkamp, Zwetsloot, & van der Wal, 2010).[9] The assumption is that these (and other) authors avoid using the term social capital because it is ambiguous and presents a potential problem of redundancy or tautology.[10] Nevertheless, this cannot be regarded as a serious obstacle to further development of the research agenda. Much more questionable is the common situation where researchers use not only different but also non-comparable instruments, sampling, and indicators to measure social capital and draw far-reaching conclusions on that basis.

Taking into account the available research findings generated within the frame of reference of social capital, it can be said that they primarily confirm the positive influence of social capital—or one of its dimensions or antecedents—on the innovation process in a regional framework.[11] This is the good news. The bad news is that in many ways they demonstrate controversial views and explanations of this influence, and we often face contradictory claims. The majority of authors are convinced that social capital "exists" as a driver of innovativeness (Tamaschke, 2003) and that it matters, but we still know very little about the ways it matters and its real effects (externalities) (see Akcomak & Ter Weel, 2009; Rutten & Gelissen, 2010).[12]

THE NATIONAL (COUNTRY) LEVEL

The authors of the report on the state of innovation activity in the EU (European Innovation Scoreboard—EIS, now IUS—Innovation Union Scoreboard) stress social capital's special role in the form of generalised trust and the perception of corruption in the analysis of factors that enable more intense innovation at the national level (Hollanders & Arundel, 2007). In this respect, the question arises of why they have chosen these two particular dimensions (indicators) of social capital (in reality, the perception of corruption has very little to do with social capital, it is an indirect and inverse indicator; see the arguments by Van Deth, 2003). According to some authors, such a simplified procedure cannot match the multi-dimensionality of this factor and consequently cannot provide reliable results (notably Doh & Acs, 2011, also see Häuberer, 2011). Nevertheless, this and other studies may have a positive effect since they stimulate other authors to test the influence and significance of social capital in the framework of innovation performance on different levels by using more complex research designs. It must be recognised that this kind of research is still in the pilot phase. The table in the Appendix shows that since 2004 only five studies on the national level have been conducted—there are perhaps more of these although Doh and Asc (2011) collected such studies and were not able to find any after 2004.

If we look at the table in the Appendix showing the national level regarding the sampling and data used, it is clear that all the studies are quantitative and based on secondary data analysis. Two of them are limited to the European context, while the rest include countries from around the world.

With regard to the measure for social capital we can see a similar set of indicators with two exceptions: Hollanders and Arundel (2007) use just two indicators, while Bruno et al. (2008) employ a composite index of the socio-cultural environment, whereby social capital is operationalised in the form of specific and specialised indicators at firm level (cooperation and contacts with stakeholders). This could be a productive exception and good example for other researchers. Unfortunately, they use inadequate items for some indicators of social capital. This can be demonstrated with the high position taken by Slovenia. Only Finland ranks higher, with Slovenia, Denmark, and Sweden being very close, while Austria, Germany, and even the Netherlands and the U.K. are much further behind. In the indicator »Cooperation with academic world« Slovenia even ranks second, with only Finland being ahead of it. This data which is taken from Eurostat (CIS-4) and stems from a survey on innovative firms provides an insufficient basis for making conclusions such as those of Bruno et al. (2008). Yet, if Slovenian experts and policymakers were asked about this cooperation, they simply would not agree, saying that this one of the most critical issues in the country's innovation system. Other surveys and rankings also considering this kind of cooperation paint an entirely different picture (for instance, the World Competitiveness Report by the IMD shows the relatively low position of Slovenia regarding cooperation between firms and the academic sphere—this information was acquired by surveying representatives of management).

As mentioned, here we are dealing with more specific and specialised measures of social capital but a certain contradiction can be observed between this alternative (and in principle more valid) measure and conventional indicators (trust, associational involvement). Namely, while Slovenia ranks very highly in the alternative index, its ranking in the conventional index is around the average or below average—depending on the data set used. It is interesting that with regard to generalised trust (which in this study—Bruno et al., 2008—forms part of the social capital index) that belongs to conventional (contextual) measures, Slovenia ranks very low—only three countries have a lower score). Furthermore, Slovenia's rankings in other segments of the socio-cultural environment utilised in the Bruno et al. study are much lower than its social-capital rankings. How can we understand this controversial situation? The only solution is to verify and extend the empirical evidence and to become familiar with nation-specific findings and discussions.

In summarising the findings about the impact of social capital, one can generally say that this type of capital matters for innovation performance on the national level. However, there are some variations and reservations. The greatest similarity is found between Doh and Acs and Dakhli and De Clercq concerning the research design, sampling, and source of data (with the only difference being the time period involved), yet this similarity is not reflected in convergent conclusions. The latter stress only a partial effect of social capital and underline that the different dimensions of social capital cannot be conceived as one integral construct, while the former state they discovered intercorrelations between all the dimensions used in their investigation. How can we explain this difference and incongruence? Can this difference be linked to the fact that the first study was conducted recently and took the dataset from the fifth round of the WVS (2005–07) into account, while the analysis of Dakhli and De Clercq is based on data from the WVS 1995 and data for innovation performance from 1998?

Especially provocative is the thesis—otherwise in line with Putnam's theoretical expectations—about the empirical congruency of main dimensions like trust, norms of reciprocity, and engagement in different types of networks (Doh and Acs, 2011; Doh and Mc Neely, 2011). It is namely known that many researchers have been unable to identify the existence of such a unified construct. For instance, there has been a long-running debate about the relationship between trust and associational involvement (several authors in Van Deth et al., 2007, also see Haynes, 2009) or between social capital (in a narrow sense) and civic norms resulting in the assertion that such an association cannot be statistically confirmed (Zmerli, 2010; Haynes, 2009). However, it is very difficult to decide who has the better arguments; in my opinion, much of this evidence is based more on 'virtual' quantitative calculations without a basic judgment or assessment of the (input) data quality, including the items and indicators used. More fundamental questions should be addressed. Even if we accept the (general) thesis that social capital matters for national innovation performance, we need to explain at least three sets of issues concerning the direct or indirect influence of social capital, its weight compared with other drivers of innovation performance, and the causal direction.

1. The studies under scrutiny do not explain whether social capital—either its individual dimensions or as an integral construct—has a direct or indirect (contextual) impact on innovation capacity. Yet it can be inferred from the selected indicators (with the exception of Bruno et al., 2008) that social capital plays more of an indirect and mediating role. It is a segment of the socio-cultural innovation milieu which may stimulate knowledge sharing or creative problem-solving within regional or national innovation systems but cannot be seen as a crucial or determining factor directly influencing innovation. This

does not mean that such an influence cannot be found, only that the empirical evidence so far is scarce and tends to support the indirect role of social capital.
2. Consequently, is it also unclear how much variance in innovation performance can be explained by the individual dimensions or integral construct of social capital. What is the 'weight' of this factor in comparison with other factors or (in)tangible capitals influencing or determining innovation performance like investment in R&D, human capital, entrepreneurship, institutional regulation mechanisms, the strategic capacity of policy, and management actors, etc.? The selected indicators and above all sampling do not allow us to conclude which factor is more significant or important.
4. What is the causal direction—is social capital the 'cause' or consequence of innovation or developmental performance? Are countries more successful because of social capital or do they have a high level of it because they are more developed and innovative? There is even a third possibility, namely, that the existence of social capital is a precondition for such success as well as its consequence in the sense of a feed-back loop (or virtuous circle). The statement that "this paper's regression analyses show a positive relationship between social capital and innovation at the country level" (Doh and Acs, 2011, 10) can be understood in many ways. Basically, this means that existing measurements are far from being precise and robust, it only indicates a '(cor)relation,' it may be statistically significant, but this significance is only founded and understood in theoretical models.

THE REGIONAL LEVEL

Of the 12 related studies (see the Appendix), only one is qualitative in character (Fromhold-Eisebith, 2004); the rest are quantitative and with one exception (Kallio et al., 2010, and partly in Bruno et al., 2008) they are all based on empirical evidence from secondary data analysis. Regarding the measures for social capital there are not many differences, the majority use multidimensional indicators (an index). A similarity with some variations (especially in the case of Dominicis et al., 2011) can also be noticed with regard to the measures used for innovation performance. Again, even where authors display similar indicators their results are inconsistent. While it is true that all the studies under scrutiny show that social capital or its individual dimensions are somewhat positively correlated to regional innovation performance, it remains unclear how strong this connection is in reality and how it operates in interaction with other innovation drivers.

Some authors dealing with the regional level also focus on trust, which in their eyes is the best proxy for social capital or even identical to it.

Their results show that trust (generalised and in some cases also institutional) is statistically and significantly connected with developmental performance or innovations (Akcomak & Ter Weel, 2009; Blume & Sack, 2008; de Clercq & Dakhli, 2004). However, most authors use associational involvement and a network approach as measures of social capital, in addition to trust.

Nonetheless, very different findings are evident since some claim that trust is an important predictor of innovation activity, while others deny this. The often quoted findings of Beugelsdijk & van Schaik (2005) on the influence of social capital in regional development present one study that does not support the thesis of the importance of trust. The same is true for some other, more specific (innovation-oriented) studies (Hauser, Tappeiner & Walde, 2007; Kaasa, 2007; Landry, Amara & Lamari, 2002). Some of them (e.g., Hauser, Tappeiner & Walde, 2007; Kaasa, 2007; Dominicis et al., 2011) point out the significance of associational involvement or civic participation and/or informal networks for innovative performance, rather than trust.

Even authors using the same dataset (EVS or ESS) come to different conclusions. The findings presented by Hauser, Tappeiner, & Walde (2007) show that trust has no significant impact on innovation performance, while a study by Rutten & Gelissen (2010) stresses the importance of trust (both used the EVS dataset). The results of the former study indicate a strong influence of associational involvement, while the latter concludes that social networks have a small but positive impact on innovation. While one study based on ESS data concluded there was a positive impact of generalised trust on innovation activities (Akcomak & Ter Weel, 2009), other research (Kaasa, 2007) based on the same dataset concludes that trust has a small effect on patenting. At first glance, the studies by Akcomak & Ter Weel (2007) and Kaasa (2007) arrive at similar conclusions; however, it is clear that the latter author identified voting as the strongest factor. In addition, they employ very different indicators (indices) of social capital.

Referring to findings on the regional level, the same questions as for the national level remain relevant and relatively unanswered: the issue of the (in)direct impact of social capital, its weight and significance in comparison with other enablers of innovation capacity, as well as the causal direction.

METHODOLOGICAL REASONS FOR THE DIVERGENT FINDINGS

One main problem which has received too little attention is the question of the frequent divergence between the findings of these studies. This is also true of studies conducted over a longer timeframe, encompassing the last 15 years, and dealing with broader aspects of regional development, not only innovation (see Westlund & Adam, 2010; for further reading on the incongruent findings, see Field, 2008, and Häuberer, 2011).[13]

To avoid any misunderstanding: the goal is not to set up a kind of unified methodology for such a contextually sensitive and multidimensional notion as social capital (innovation activity, in contrast, is much easier to measure). However, the incongruent or even contradictory results do not allow an adequate interpretation and comparison. In my opinion, these divergent findings and unilateral interpretations are a consequence of the following methodological shortcomings.

1) Unselected *quantification* is noticeable.[14] With one (small) exception, the large majority of the reviewed studies (17) are quantitatively oriented and mostly use survey data or (census) data from statistical offices or Eurostat (keeping in mind that these data are also gathered through survey questionnaires) or from national surveys (like Blume & Sack, 2008; also see Freitag & Traunmueller, 2008). Furthermore, very little effort is devoted to the triangulation and utilisation of an array of alternative methods or techniques like the Comparative Qualitative Method (CQM) or fuzzy-set analysis (Ragin & Rihoux, 2009) as well as quasi-experimental design with control and experimental groups.

2) Insufficient attention is paid to the control of *quality (and credibility) of input data*. Researchers rely on a single (secondary) dataset, most often on EVS–WVS, ESS, and EB or (more rarely) on data they have acquired themselves (or collected from various sources). The obvious tendency for data and its routine quantification to be taken for granted is the key issue indicating the domination of the positivist/empiricist paradigm.

3) *Some variations in sampling and level of aggregation are not considered when generalising findings.* Certain authors, for example, take all or a majority of regions (NUTS 1 or NUTS 2) into account, while others use selected samples without explanation (typical in, for example, Hauser, Tappeiner, & Walde, 2007, where regions of only six EU countries are taken into account, excluding Nordic and other regions). This may have a great (biased) impact on the results of the analysis. In other words, if a less selective sample had been used the results might have been quite different. Variations in sampling are not problematic per se; the problem lies in the interpretation and generalisation of findings that do not take account of sampling limitations and peculiarities. Sometimes the level of aggregation (societal or individual) is also not taken into account, which can lead to a so-called ecological fallacy—mistakenly inferring from the macro to the micro level. Correlations that explain the differences between countries might not explain differences between individuals within a country (Welzel & Inglehart, 2010).

4) *Various indicators (operationalisations) of social capital* are utilised (sometimes also for innovation activity). Some authors aim to encompass all three main dimensions (trust and other norms, associational

involvement and informal networks in their various forms: weak or strong ties, informal contacts etc.), while others are more selective or arbitrarily satisfied with one indicator or proxy (most often trust).[15] Not much effort is made to apply more specific indicators (an exception on the national level is Bruno et al., 2008, and Kallio et al., 2010, on the regional level).

5) Unusual or *artificial (mechanical) correlations* arise at times. One study shows a high level of correlation between innovation activity and voting behaviour, which is treated as civic participation (Kaasa, 2007). What high voting participation has to do with a propensity for innovation seems to be a quite trivial question. This is more a statistical artefact than a serious and conceptually relevant correlation. Is quite common that (some) researchers try to avoid searching for reasons and additional re-testing (falsification) when they obtain 'strange' correlations or clusters of countries.

6) An *external and contextual view of regional social capital* predominates. The unit of inquiry is the whole region or country rather than the regional and national innovation system itself and its actors (with the exception of Fromhold-Eisebith, 2004, and Kallio, Harmaakorpi, & Pihkala, 2010; and partly Bruno et al., 2008).

HOW TO IMPROVE THE RESEARCH DESIGN

To achieve a higher degree of convergence and to increase the quality of data and the credibility of interpretations, it is necessary to pay stronger attention to *epistemic reflection* and more in-depth and synthetic approaches. More fine-grained analyses are needed to explain the modalities and mechanisms through which social capital to a greater or lesser extent positively or ambivalently affects (and stimulates) innovation performance and creative entrepreneurship.

In this regard, the following conclusions and instructions relevant to more complex research strategies should be stressed:

1) Greater emphasis should be placed on qualitative research, and especially on triangulation, that is, the combination of quantitative and qualitative methods and data sources. Most of the studies under scrutiny are based on a quantitative analysis. It would be desirable to complement and compare the results of such an approach, for instance, with conclusions deriving from focus groups (or semi-structured interviews). However, we have to bear in mind that in-depth case studies (a case-based approach) are quite realistic on the micro (local) level.

2) On the macro level—especially when dealing with secondary data—a different strategy should be employed, namely, meta-analysis should

be utilised as a tool to verify the validity of findings and the quality of data by comparing different sources and datasets (examples can be found in Westlund and Adam, 2010; Zheng, 2008). The first and simplest step in this direction would be a systematic comparison of cross-national datasets and findings acquired by (domestic) national (regional) studies.

3) It could be fruitful to shift from general sampling (which only provides contextual information) to more specific sampling consisting of subgroups (*target groups*) deemed to play an eminent role in regional and national innovation activities (see Rutten & Gelissen, 2010). In addition, greater stress should be placed *on insiders' (inductive) observations* focussing on actors in regional innovation systems—such as R&D personnel, independent inventors, representatives of civil society initiatives, intermediary or supporting institutions, and policymakers—by using methods including surveys, semi-structured interviews, critical discourse analysis, and qualitative network analysis (combined with quantitative analysis).

4) Following the change in sampling, new more specialised and specific indicators of social capital should be formulated and utilised, including the share of people with experience in teamwork and the project type of organisation, the density of 'industrial districts,' contacts between firms and academic institutions, and similar types of collaborative structures.

SOME CONCLUDING REMARKS: TOWARD AN EPISTEMIC TURN

It is very difficult to speak of any new insights or methodological improvements arising from this analysis. It is true that the findings show social capital has a certain (greater or lesser) effect on regional innovation performance. However, there is no consensus on which dimensions are relevant, and little is known of the mechanisms through which social capital is transformed into the catalyst and facilitator of innovation activity and a creative milieu. One of the findings and conclusions from the meta-analysis relates to the distinction between contextual or generalised social capital and specialised social capital. The indicators for the former are connected with associational involvement and civic norms (including trust) as well as with primordial social capital (informal ties and egocentric networks). It is hypothesised that this form of social capital which characterises a region or country positively influences the emergence of a specialised form of social capital—and vice versa. For the latter form of specialised social capital, another set of indicators has pointed to cooperation in teamwork or project organisation or to the development of collaborative structures among firms (industrial districts or clusters) or between them and other institutions like universities or supporting institutions within regional or national innovation systems. However, such a trade-off (or even spillover) between both forms/levels is theoretically justified but difficult to prove since it requires a complex methodological approach. The available fragmentary evidence suggests it is also possible to have weak generalised and strong specialised social capital (such as the case of Slovenia in Bruno et al., 2008).

The other issue concerns the unresolved problems relating to the epistemic (and ontological) status of social capital as a contested concept as well as the repetition of the same methodological shortcomings. A possible consequence is that many researchers are aware and tired of these continuing unsettled questions summarised in the syntagm of an essentially contested concept. They may be trying to escape by focusing on the dimensions of social capital without evoking the actual term. However, some insist on looking for new solutions under the umbrella of social capital. Both groups can be productive in the presence of certain preconditions, some of which have been presented in this article. It is especially important for empirical research to be concentrated on the concrete actors and institutions in the regional innovation system, which presupposes the utilisation of multiple methods and sources, and greater attention being paid to data quality.

It seems that only a radical modification of research design and methodological thought can give a fresh impetus and allow us to gain a deeper understanding of social capital's role and/or its dimensions in (regional) innovation processes. Dialogue and the exchange of ideas among researchers are inconsistent with the routinisation and uncritical reception of the available research findings. However, this does not mean these findings should be denied or that we need to start from scratch. In terms of the existing body of knowledge, the meta-analysis approach is appropriate in order to discern the most relevant, grounded, and comparable theses and data. It is also important to more extensively exploit the available datasets by combining variable-based and case-based approaches or, put in more operative terms, by combining cross-national findings and national studies in order to achieve more valid knowledge. In the case of new research, more attention should be paid to the ethnographic qualitative approach and to triangulation.

My thesis is that this change in research methodology denotes a shift from the positivist/empiricist paradigm to the post-positivist/post-empiricist model of knowledge and data production.[17] Here we are speaking about an "epistemic turn" which affects not only social capital research but the entire social science research design which should become more complex and more reflexive. This shift/transition is already underway not as a radical breakthrough but as an incremental evolution involving many small steps. One of these would be to organise the research process differently with an accent on a more coordinated approach at the European level, while a second would involve the systematic comparison of cross-national surveys with findings made at the national level.

Appendix 5.1 Summary Overview of Studies Dealing with the Impact of Social Capital on National and Regional Innovative Performance

Studies/ Publication	Sampling/ Source	Measures of Social Capital	Measures of Innovation Performance	Impact of Social Capital
Dakhli & De Clercq (2004) (article)	59 countries from all continents (Quantitative) SDA aggregate sampling across (not within) countries	Gen. and instit. trust; associat. involvement; norms of civic behaviour from WVS 1995	Patents; R&D; high-tech export from World Bank 1998	Partial positive impact of trust and assoc. involvement. Weak or negative for norms of civic behaviour
Hollanders & Arundel (2007) (working paper)	EU-27 (Quantitative) SDA	Gen. trust; perception of corruption	composite EIS index	Positive and strong
Willems (2007) (master's thesis)	43 countries (Quantitative) SDA	Bridging SC*: assoc. involvement Bonding SC: significance of family and friends data from WVS 1995	R&D exp. (UNESCO) Patents per country USPTO/ EPO	Both types positive sign. (both types inter-correlated)
Bruno et al. (2008) (working paper)	EU-25 (Quantitative) SDA	SC included in index of socio-cultural environment** SC: types of a firm's cooperation partners; from various sources EUROSTAT, EB, EVS etc.	EPO patents BERD	Positive: the weaker the socio-cultural environment, the lower the number of patents (and vice versa), (especially true for BERD)
Doh & Acs (2011) (article)	53 countries (Quantitative) SDA	1. Assoc. involvement 2. Trust 3. Norms of civic behaviour From WVS 2005–2007	USPT patents R&D intensity, entrepreneurship from GEI***	Positive sign. – all the dimensions are inter-correlated SC as one construct
Akcomak & Ter Weel (2007) (working paper)	83 EU regions 1990–2002; (Quantitative) SDA	gen. trust plus SC index (altruism, fairness, attitude to voluntary orgs and participation (data from ESS)	Innovation index, annual GDP/ capital growth, human capital plus success of EU support programmes	SC positive (sig.) impact on innovation; positive correlation between SC and EU support for regional innovation

Continued

Appendix 5.1 Continued

Studies/ Publication	Sampling/ Source	Measures of Social Capital	Measures of Innovation Performance	Impact of Social Capital
Akcomak & Ter Weel (2009) (article)	102 EU regions; quantitative SDA	Gen. trust (data from ESS)	R&D intensity, patent EPO applications, human capital (plus path dependency)	On growth insignificant, on innovation positive, trust has an impact on innov. and innov. on growth
Barrutia & Echeberria (2010) (article)	Italian and Spanish regions NUTS I/ 2; quantitative SDA	(1) Rational choice – own proxies; (2) sociological approach in the form of the Beugelsdijk index (EVS)	Patents EPO per million, R&D expenditure	Trust and B. index less important, rational choice more important [16]
Blume & Sack (2008) (article)	74 German (West) Regions; (Quantitative) SDA	Post-materialist values, civic and political networks; domestic data sources (data from Allbus, Forca, Soep)	Annual growth in output per worker, patenting is included but as an independent variable and is not discussed separately in the text	Regarding productivity: assoc. involvement no impact, while trust has a positive correlation Regarding patents (1992–1994): neither trust nor civic assoc. have any significant impact on patents
Hauser, Tappeiner, & Walde (2007) (article)	51 regions from 6 EU countries; quantitative SDA	Trust, assoc. invol., weak and strong ties (EVS)	Patenting; human capital	Active membership strong impact, trust not important, weak ties important

Studies/ Publication	Sampling/ Source	Measures of Social Capital	Measures of Innovation Performance	Impact of Social Capital
Kaasa (2007) (article)	162 EU regions; (Quantitative) SDA	6 factors, 20 indicators of SC (trust, assoc. involv., civic participation (voting) – ESS)	Human capital, R&D exp., patenting	Dissimilar effects: trust a small effect on patenting, voting the strongest influence
Kallio, Harmaakorpi, & Pikhala (2010) (article)	Case study – Lahti region (Finland), own (quantitative survey) data, respondents: actors of RIS (N = 234)	Personal SC – brokerage	Absorptive capacity	Gatekeepers (brokers) with bridging as well as bonding CS increase knowledge transfers (in the region without intensive R&D activity)
Rutten & Gelissen (2010) (article)	120 EU regions in EU-15 (mix of NUTS I/2); Quantitative SDA	Assoc. involvement, i.e. social networks, values (self-expression, tolerance, Protestant ethic) – EVS data	Patents, human capital (plus indicators of wealth or economic development)	Self-expression (trust) the strongest impact on innovation, social networks a small positive effect on wealth via innovation
Fromhold-Eisebith (2004) (article)	Aachen planning region, own (qualitative) data from interviews and other secondary data	Informal contacts among actors of RIS	Increase in number of hi-tech firms	Positive, significant

Continued

Appendix 5.1 Continued

Studies/ Publication	Sampling/ Source	Measures of Social Capital	Measures of Innovation Performance	Impact of Social Capital
Laursen et al. (2011) (article)	2,400 firms in 21 Italian regions Quantitative SDA	10 variables like friendship, spare time participation in (non)voluntary organisations and number of voluntary orgs. per region, political activity and donation	R&D exp. External R&D from Capitalia 2006 and EUROSTAT 1999	Positive; regions with a higher level of SC are most likely to introduce product innovation – the same holds for R&D in firms
Mascierelli (2011) (book)	4,735 firms 21 Italian regions (Quantitative) SDA On the basis of Unicredit	Active participation in selected voluntary and involuntary organisations; relations with friends****	Products and process innovations in firms	High SC settings have a positive impact on regional creativity and both on innovation success
Dominicis et al. (2011) (working paper)	146 NUTS 2 EU regions (in 11 countries) (Quantitative) SDA	SC-1 = a level of education, % older than 65; share of persons reading newspapers, life satisfaction (from EB 1999–2001) SC-2: trust, from EB 1996	Patents EPO R&D intensity employment in hi-tech firms. from EUROSTAT 2000-2002	SC-1: positive, trust negative (insignif.)

Key: ESS = European Social Survey; EVS = European Values Study; RIS = Regional Innovation System; SC = Social Capital; SDA = Secondary Data Analysis; BERD = Business Expenditure for R&D; EB = Eurobarometer.
*The study includes Hofstede's cultural dimensions as an additional intervening variable.
**Socio-cultural environment is composed of organisational, human, cultural, and social capital.
***In this study, control variables such as unemployment rate, income gap (Gini), and country size are also considered.
****The author uses the notion of creativity or human capital and cultural capital as additional or intervening variables. She also presents other studies relating to firm-banking relationship as well as firms-offshore relationship.

NOTES

1. Parts of this text are partialy based on the article Adam, F. (2011), Regional Innovation Performance in Light of Social Capital Research and Applicaton, Social science Information, 50 (3–4) 414–428.
2. One concept that overlaps with social capital is social cohesion. According to some authors, we are dealing with a framing concept. Social cohesion is a 'framing concept' an overarching notion that links different policy areas and responds to the growing need in policy discourse for integrative models that help make sense of issues as diverse as income security, neighbourhood safety and housing" (Toy, 2006).
3. Speaking in linguistic terms, it can be argued that the social-capital concept is a strong signifier without a signified object. Empirical findings are also known not to allow the conclusion that this concept is a broader construct composed of dimensions (variables) which are inter-correlated (as, for instance, Putnam's approach presupposes). Most researchers report they have been unable to find a connection between trust and associational involvement or that there is no correlation between values (norms) like reciprocity and other (structural) dimensions of social capital (see Van Deth, Montero, & Westholm, 2007). It is true that these findings may not be adequate; nevertheless, this is the current situation. There are ongoing efforts to ensure a theoretical foundation (Fulkerson & Thompson, 2008; Häuberer, 2011).
4. A similar position has been formulated as follows: "We want to maintain the original idea of social capital as a species of capital, and believe that the strength of the concept lies just in the recognition of this capital nature" (Tura & Harmaakorpi, 2003: 10). On the other side, some authors argue that social capital is not a proper 'capital' comparable with economic or human capital and that social capital should not belong to the 'capital family' (Westlund, 2006; Hanley, 2009). Nevertheless, they still use this term because it is widely accepted.
5. GREMI is an acronym for a French 'school' focused on regional innovation research (Groupe de Recherche Européen sur les Milieux Innovateurs), which has been active since the mid-1980s.
6. The position of the author of 'creative class,' R. Florida, is also paradoxical in this regard. He (wrongly) states that social capital refers (exclusively) to closed networks and strong ties, and is therefore unable to capture network diversity and tolerance toward minority groups as key factors of a creative milieu. As a consequence, he rejects the use of this term (Florida, 2005; Florida et al., 2002). It seems that some authors even problematise those aspects which have already been resolved (as with the distinction between bonding and bridging social capital).
7. The same author asks: "What is the relationship between social network and social capital?" as one of the "many unanswered questions." Another author considers trust and social capital to be two different entities. He concludes that "social capital may be the force that has kept trust low" (Keele, 2007).
8. It is interesting that Woolcock himself in one very long sentence uses both labels for social capital, pro-theoretical and framing concept (Woolcock, 2010, 474).
9. Some authors used the concept of social capital in their early works when they elaborated the types of networks most suited for innovation activities (Landry, Amara, & Lamari, 2002). Later on, the same authors only use terminology specific to network theory, without the term social capital being mentioned at all (Ouimet, Landry, & Amara, 2007).

10. Some Italian authors (economists) propose—in order to avoid the vagueness of the social- capital concept—using *civic capital* instead of social capital. Civic capital is defined as "persistent and shared beliefs and values that help a group to overcome the free rider problem" (Guiso, Sapienza, & Zingales, 2010, 7). They exclude networks and insist that values and trust are the essence of civic capital.
11. A partial exception is the study of Italian and Spanish regions where the authors used two sets of indicators (indexes) for social capital, the first being on rational-choice theory and the second, the so-called sociological index, based on the Beugelsdijk & van Schaik approach. While the latter did not show any substantial impact of social capital on innovation outcomes, the rational-choice-driven measurement pointed to social capital's high relevance to innovation performance (Barrutia & Etchebarria, 2010).
12. The latter authors conclude: "The exact transformation of social capital into innovation remains unclear" (Akcomak & Ter Weel, 2009).
13. A statement from an older OECD study remains valid: "As in the case of human capital, the evidence is affected by the quality and breadth of proxy measures, the complexity of inter-relationships between different conditioning factors, and the difficulty in comparing countries with widely differing cultural, institutional and historical conditions" (OECD, 2001).
14. Relating to the measurement of social capital, one of the most prominent sociologists observed as follows: "In spite of its theoretical weakness, [social capital] has outlasted its competitors partly because no other can match the catchiness of the word and partly because it seemed to promise the possibility of quantification in an analogy with its economic inspiration. However, this promise did not materialize. The method used in studying those aspects of social reality that are covered by this term have proved less adequate and the attempt at quantification illusory" (Luckmann, 2007, 10).
15. The problem of (construct) validity is critical in these studies, and the following warning is instructive: "First, the operationalizations of our variables do not satisfactorily measure the underlying theoretical concepts" (Rutten & Gelissen, 2010, 936).
16. The authors use so-called rational-choice-driven metrics. However, their proxies—such as unemployment rate, life expectancy Gini index, loans/GDP—for social capital are problematic at first sight. For instance, for the variable "rate of survival of social capital stock" they use "unemployment rate" as a proxy, etc.
17. A postpositivist or postempiricist approach points to a deliberative way of dealing with data and methods for collection and processing of data. It cannot be understood as an antipositivism since it is interested in scientific method (including quantitative methods). However, its accent is on meta-analysis, triangulation, systematic falsification procedure, and critical discourse between scientists. In the literature is not yet elaborated in systematic form, though some authors represent it in a way which is close to my understanding (see for instance, Fischer, 1998).

REFERENCES

Ahrweiler, P. (ed.). (2010) *Innovation in complex social systems.* London: Routledge.

Akcomak, S., & Ter Weel, B. (2007) How do social capital and government support affect innovation and growth? Evidence from the EU regional support programmes. UNU-MERIT Working Paper Series # 2007–009, United Nations University. *http://ccp.merit.unu.edu/2007/3140*. Accessed March 19, 2010.

Akcomak, S., & Ter Weel, B. (2009) Social capital, innovation and growth: Evidence from Europe. *European Economic Review, 53* (5), 544–567.

Asheim, B.T., Coenen, L., & Svensson-Henning, M. (2003) *Nordic SMEs and regional Innovation systems: Final report*. Lund: Lund University. http://www.nordicinnovation.net/_img/smes_and_regional_innovation_systems.pdf. Accessed May 15, 2010.

Barrutia, J.M., & Echebarria, C. (2010) Social capital, research and development, and innovation: An empirical analysis of Spanish and Italian regions. *European Urban and Regional Studies*, 17(4), 371–385.

Beugelsdijk, S., & van Schaik, T. (2005) Social capital and growth in European regions: An empirical test. *European Journal of Political Economy, 21*, 301–324.

Blume, L., & Sack, D. (2008) Patterns of social capital in West German regions. *European Urban and Regional Studies, 19*, 229–248.

Boschma, R.A. (2005) Proximity and innovation: A critical assessment. *Regional Studies,* 38 (1), 61–74.

Bruno, N., Miedzinski, M., Reid, A., & Ruiz Yaniz, M. (2008) Socio-cultural determinants of innovation, Technopolis (WP 10). Europe Innova.: *http://www.technopolis-group.com/resources/downloads/Socio-cultural-factors-innovation.pdf*. Accessed December, 2011.

Cooke, P. (2008) Regional innovation systems, clean technology & Jacobian cluster-platform policies. *Regional Science Policy & Practice,* 1 (1), 23–45.

Costa, A.C., Bijlsma-Frankema, K., &de Jong, B. (2009) The role of social capital on trust development and dynamics: Implications for cooperation, monitoring and team performance. *Social Science Information,* 48 (2), 199–228.

Dakhli, M.,& De Clercq, D. (2004) Human capital, social capital, and innovation: A multi-country study. *Entrepreneurship & Regional Development: An International Journal,* 16 (2), 107–128.

Doh, S., & Acs, Z. (2011) Innovation and social capital: A cross-country analysis investigation. *Industry&Innovation,* 17 (3), 241–262.

Doh, S., & Mc Neely, C. (2011) A multi-dimensional perspective on social capital and economic development: An exploratory analysis. *The Annals of Regional Science (DOI 10. 1007/s00168–011–0449–1)*.

Dominics, L., Florax, J.G.M.R., & de Groot, H.L.F. (2011) Regional clusters of innovative activity in Europe: Are social capital and geographical proximity the key determinants? *Tinbergen Institute discussion paper* (TI 2011–009/3). Rotterdam: Tinbergen Institute.

Field, J. (2008) *Social capital,* 2nd edition. London: Routledge.

Fischer, F. (1998) Beyond empiricism: Policy inquiry in postpositivist perspective. *Policy Studies Journal,*26 (1): 129–146.

Fleming, L., Mingo, S., & Chen, D. (2007) Collaborative brokerage, generative creativity and creative success. *Administrative Science Quarterly,* 52, 443–475.

Florida, R. (2005) *The flight of the creative class: The new global competition for talent*. New York: HarperBusiness.

Florida, R., Cushing, R., & Gates, G. (2002) When social capital stifles innovation. *Harvard Business Review. http://hbr.org/2002/08/when-social-capital-stifles-innovation/ar/1*. Accessed May, 2011.

Freitag, M., & Traunmueller, R. (2008) Sozialkapitalwelten in Deutschland. *Zeitschrift für vergleichende Politikwissenschaft,* 2, 221–256.

Fromhold-Eisebith, M. (2002) Innovative milieu and social capital: Exploring conceptual complementarities. *Conference report, 42nd Congress of the European*

Regional Science Association (ERSA). Dortmund, Germany, August 27–31, 2002.
Fromhold-Eisebith, M. (2004) Innovative milieu and social capital: Complementary or redundant concepts of collaboration-based regional development? *European Planning Studies* 12 (6), 747–765.
Fulkerson, G.M, & Thompson, G. (2008) The evolution of a contested concept: A meta-analysis of social capital definitions and trends (1988–2006). *Sociological Inquiry*. 78 (4), 536–557.
Guiso, L., Sapienza, P., & Zingales, L. (2010) Civic capital as the missing link. http://www.kellogg.northwestern.edu/faculty/sapienza/htm/civic_cap.pdf. Accessed June 20, 2010.
Häuberer, J. (2011) Social capital theory: Towards a methodological foundation. Wiesbaden: Springer Fachmedien.
Hauser, C., Tappeiner, G., & Walde, J. (2007) The learning region: The impact of social capital and weak ties on innovation. *Regional Studies*, 41 (1), 75–88.
Haynes, P. (2009) Before going any further with social capital: Eight criticisms to address. *Ingenio Working Paper*: 2009/02. http://digital.csic.es/bitstream/10261/14203/1/Before_Going_Any_Further_With_Social_Capital__Eight_Key_Criticisms_to_Address%5B1%5D.pdf. Accessed March, 2012.
Hollanders, H., & Arundel, A. (2007) Differences in socio-economic conditions and regulatory environment: Explaining variations in national innovation performance and policy implications. *INNO Metrics—Thematic Papers*. Maastricht: Maastricht University.
Isaksen, A. (2009) Innovation dynamics of global competitive regional clusters: The case of Norwegian centres of expertise. *Regional Studies*, 43 (9), 1155–1166.
Kaasa, A. (2007) Effects of different dimensions of social capital on innovation activity: Evidence from Europe at the regional level. *Technovation*, 29 (3), 218–233.
Kallio, A., Harmaakorpi, V., & Pihkala, T. (2010) Absorptive capacity and social capital in regional innovation systems: The case of the Lahti region in Finland. *Urban Studies*, 47 (2), 303–319.
Keele, L. (2007) Social capital and the dynamics of trust in government. *American Journal of Political Science* 51 (2), 241–254.
Landry, R., Amara, N., & Lamari, M. (2002) Does social capital determine innovation? To what extent? *Technological Forecasting and Social Change*, 69 (7), 681–701.
Laursen, K., Masciarelli, F., & Perncipe, A. (2011) Regions matter: How localized social capital affects innovation and external knowledge acquisition. *Organisation science orsc.1110.0650*; published online before print. Accessed May 17, 2011.
Liebowitz, J. (2007). Social networking: The essence of innovation. Lanham, MD: Scarecrow Press.
Luckmann, T. (2007) Preface. In Adam, F. (ed.), Social capital and governance. Berlin, London: LIT Verlag.
Masciarelli, F. (2011) The strategic value of social capital. Cheltenham: Edward Elgar.
Nijkamp, P., Zwetsloot, F., & van der Wal, S. (2010) Innovation and growth potentials of European regions: A meta-multicriteria analysis. *European Planning Studies*, 18 (4), 595–611.
OECD (2001) The well-being of nations: The role of human and social capital. http://www.oecd.org/dataoecd/30/40/33703702.pdf. Accessed November, 2011.
Ouimet, M., Landry, R., &Amara, N. (2007) Network positions and radical innovation: A social network analysis of the Quebec Optics and Photonics Cluster. *International Journal of Entrepreneurship and Innovation Management*, 7 (2), 251–271.
Ragin, C., & Amoroso, L. (2011) *Constructing social research*, 2nd ed. Los Angeles: Sage.

Ragin, C., & Rihoux, B. (2009) Configurational comparative methods: Qualitative comparative analysis and related techniques. London: Sage.

Russo, M., & Rossi, F. (2008) Cooperation networks and innovation: A complex system perspective to the analysis and evaluation of an EU regional innovation policy programme. *MPRA paper no. 10156.*

Rutten, R., & Gelissen, J. (2010) Social values and the economic development of regions. *European Planning Studies*, 18 (6), 921–939.

Tamaschke, L. (2003) The role of social capital in regional technological innovation: Seeing both the wood and the trees. In Huysman, M., et al. (eds.), *Communities and technologies*. Deventer: Kluwer, 241–264.

Toy, M. (2006) *Social cohesion—The Canadian urban context*. Parliamentary Information and Research Service. *http://www.2parl.gc.ca/content/lop/researchpublications/prb 0756-e.pdf.* Accessed January, 2011.

Tura, T., & Harmaakorpi, V. (2003) Social capital in building regional innovative capability: A theoretical and conceptual assessment. *Paper presented at the 43rd Congress of European Regional Science Association (ERSA)*, Jyväskylä, Finland, 27th-30th August, pp. 27–31.

Van Deth, J.W. (2003) Measuring social capital: Orthodoxies and continuing controversies. *International Journal of Social Research Methodology*, 6 (1), 79–92.

Van Deth, J.W., Montero, J.R., & Westholm, A. (2007) *Citizenship and involvement in European democracies*. Abingdon, New York: Routledge.

Welzel, R., & Inglehart, C. (2010) Agency, values, and well-being: A human development model. *Social Indicators Research*, 97 (1), 43–63.

Westlund, H. (2006) *Social capital in the knowledge economy*. Theory and empirics. Berlin, Heidelberg: Springer Verlag.

Westlund, H., & Adam, F. (2010) Social capital and economic performance: A meta-analysis of 65 studies. *European Planning Studies*, 18 (6), 893–919.

Willems, M.J.T. (2007) The influence of social capital and cultural dimensions on innovation. *Master thesis*. Maastricht: University of Maastricht.

Woolcock, M. (2010) The rise and routinization of social capital, 1988–2008. *Annual Review of Political Science*, 13, 469–487. *http://academic.live.come.* Accessed January, 2011.

Yin, C.Y., &Chiang, K.J. (2010) Bibliometric analysis of social capital research during 1956 to 2008. *Journal of Convergence Information Technology*, 5 (2), 124–132.

Zheng, W. (2008) A social capital perspective of innovation from individuals to nations: Where is empirical literature directing us? *International Journal of Management Reviews*, 12 (2), 151–181.

Zmerli, S. (2010) Social capital and norms of citizenship: An ambiguous relationship? *American Behavioral Scientist*, 53 (5), 657

6 Collaboration in Innovation Systems and the Significance of Social Capital

Hans Westlund and Yuheng Li

INTRODUCTION

What is an innovation system? Expressed in simple terms, an innovation system consists of a number of actors and the links between them, which by some type of collaboration achieve innovations, that can be defined as new products or processes that are being produced or applied for commercial or other use.[1] The actors that according to the theories usually "inhabit" an innovation system are firms, universities, and other bodies for research and education, the public sector, banks, and other credit institutions, and sometimes also civil society.[2]

In this respect innovation systems have existed long before the concept as such was invented. The internationally most well-known "innovation system" of the 19th century is what nowadays is denominated the "German national innovation system" (see, e.g., Grupp, 1998) in which the state played a decisive role.

The concept of innovation systems appears to have been coined when Freeman (1987) published his book on what he came to call the Japanese innovation system. The following year Lundvall (1988) published his first essay on national innovation systems. Since then, the academic and policy-related literatures on national, regional, technological, and sectoral innovation systems have literally exploded. If we add that the concept of innovation systems has strong similarities with concepts such as industrial districts, clusters, and triple-helix, it becomes clear that within this group practically all of the more applied theories that guide growth policies around the world today are connected.[3] The post-war Keynesian-dominated growth research at the macro level has been replaced by growth research that focuses on the meso and micro levels and the linkages between the two. The same can be said about policies: the equal sign between fiscal policy and growth policy, on the one hand, and intervention policies from above, on the other, has been replaced by a business-oriented growth policy focused on supporting the processes from below.

What then are the explanations for such a transformation of research and policy on growth? The basic rationale is, of course, that the economy in the more developed world was transformed from being dominated by manufacturing industries based on mechanical technology to a knowledge and service economy based on digital technology and on worker-consumer interaction. The extent of this still ongoing transformation is not yet possible to predict, but it has already led to major changes in people's lives, both at work and at leisure. If we limited ourselves to innovation, there are strong arguments that post-war industrial growth was based on "old" industrial innovations that enhanced the effectiveness of large companies until the firms could no longer squeeze more out of the innovations. When the growth potential of the old innovations faded out, the industrial crisis of the 1970s and 80s came along. The new knowledge economy already began to emerge and expand during the 1980s in some sectors and has since spread more and more. There are clear signs that the emergence of the knowledge economy is based on some fundamental innovations (e.g., the semiconductor and all its applications within information and telecommunication technologies), which in turn gave rise to vast amounts of resulting innovations across the economy.

The concepts of *institutions, learning,* and *collaboration* are important keys in the discourse on innovation. Various students (or scholars) of innovation research have somewhat varying emphases on different concepts of innovation systems. Edquist (1997) puts more emphasis on the institutions that govern the game and regulating existing or potential collaboration between the actors (organizations). Institutions may consist of both formal and informal rules. The "right" kind of institutions can promote collaboration in innovation systems, while the "wrong" kind of institutions hinders or prevents collusion. Lundvall (1992) places more emphasis on the cognitive processes that result in innovations, i.e., on the interactive learning and collaboration processes taking place in innovation systems. These processes consist of *production* of knowledge, *transfer* of knowledge, and *application* of knowledge—with the interaction between the three. The two perspectives should not in any way be seen as competing, but complementary. The "right" institutions facilitate learning and collaboration processes, which in turn can help to change institutions so that they even better support learning and collaboration in the innovation system.

Many forms of routine collaboration and exchange of goods and products work well with no real need of taking care of social relations. However, if there is the question of exchange and collaboration on human resources such as knowledge—and not least in terms of combining existing knowledge with new knowledge—then issues of *social capital* may play a crucial role in the success or failure.

This is a conclusion reached by among others Maskell (2000) and Westlund (2006) and highlighted in several contexts within the OECD. One example is a paper by McDaniel (2006, 16) at an OECD conference on science, technology, and innovation indicators, where she concludes that "lack of innovation is not caused by lack of technology or willingness to innovate, but rather by systemic social factors in universities that discourage research findings and ideas from being developed into innovations.... Social factors may matter far more to innovation capacities of universities than previously thought."

To take universities as an example, social factors such as norms, values, and existing networks play in many ways a central role in advancing knowledge, collaboration, knowledge sharing, and innovation. On the one hand, the research and policy questions are whether the dominant norms and values encourage the development of knowledge for innovation and collaboration and exchange of knowledge with commercial entities. In contrast to technical colleges, the university's traditional approach has been intra-academic, and collaboration with partners outside academia has actually been discouraged. Although pressure from governments and funding agencies has led to new incentives for external collaboration, it remains the case that academic research is by far the most valued factor in universities. The strong intra-academic networks have created and maintained social capital that has not been conducive to external collaboration and (commercial) innovations, and these well established networks and values do not change quickly.

On the other hand, social capital plays an important role in the collaboration between innovation agents in the system. If cooperation between players works well, if they trust one another, if they share their basic values and "speak the same language," it facilitates knowledge collaboration. This means that the social relations that the participating actors set up and nourish have an impact on how much tacit knowledge is exchanged, how quickly and effectively knowledge is processed and converted into new forms, its quality, and how quickly and effectively the tacit knowledge can result in commercial innovations.

The aim of this chapter is to categorize the various dimensions of collaboration in innovation systems and the empirical variables by which collaboration can be expressed, and to identify the collaboration processes in which social capital is important.

KNOWLEDGE PRODUCTION, COLLABORATION, LEARNING, AND INNOVATIONS

The emergence of the knowledge economy has led to a changed view of the processes of innovation and the role of knowledge in them. World War II military needs in the United States and in other combatants and nuclear research yielded strong evidence for science's immediate social benefit. Also, today's information and communications technology has many of its roots in military projects. During the 1950s and 1960s these advances contributed to a simple view of the relationship between research and economic development: if the universities received funding and carte blanche, they were expected to produce new basic knowledge that was then transferred to laboratories at institutes and companies. The latter developed inventions and prototypes that could be commercialized and mass produced, thus contributing to economic growth. This model has subsequently come to be called *linear*. Basic research would be a prime mover in social development, but the university's own role was only to supply basic knowledge. Applied research, innovation, commercialization, and mass production were the continuing links of the chain, but were accomplished by other actors.

A fundamental criticism of the linear model and the universities' traditional role has been introduced by Gibbons et al. (1994) and Nowotny et al. (2001). Their thesis is that universities' role in society's knowledge production has gone from a "Mode 1" to a "Mode 2." According to this approach, Mode 1 represents the traditional intra-scientific knowledge production universities were mainly engaged in, while Mode 2 describes the integration of knowledge production with the rest of society that characterizes what we here call the knowledge economy. In Mode 2 the production of knowledge increasingly occurs through collaboration, not only between academic disciplines but also with actors outside academia, i.e., research users that determine the relevance of the new knowledge and thus are involved in quality control. Another characteristic is that a number of players other than universities contribute to the development of knowledge and collaboration with research users. The role of universities in Mode 2's knowledge production is thus not obvious.[4] In this context, it should of course be noted that this approach has met with strong criticism from defenders of the traditional academic independence, who argue among other things that the role of universities as independent truth-tellers is threatened if they are reduced to being a player in regional or national industrial policy.

Since the 1980s, the linear model has been subject to extensive criticism, and alternative approaches and models, based on the abovementioned theories of industrial districts, clusters, and innovation systems, have been launched and have found applications in industrial policy in many countries. The difference between the traditional, linear approach and the newer interactive approach that the theories of innovation systems are part of is illustrated in Figure 6.1. The figure can be used to illustrate the distinction between the two approaches with regard to key concepts of learning and institutions. In the traditional approach, learning takes place partly within each sphere of influence, and partly through the transfer of formal (codified) knowledge "from left to right" in the figure. Within each sphere learning occurs through the transfer and processing of both "silent" (tacit) knowledge and codified knowledge and feedback between the "sender" and "receivers." Between the spheres, in the linear model usually only codified knowledge is transferred and feedback mechanisms are lacking, while in contrast in the interactive model there is transfer and learning of both codified and tacit knowledge both within and between spheres.

The linear model is thus based on the premise that the knowledge being transferred from the academy already is formalized, written down, and packaged for delivery to commercial users. In the pure world of the linear model no "silent," informal (tacit) knowledge seeps from the academy before it has been formalized and there is no collaboration with users, in addition to the handover. In the interactive model there is a continuous transmission and feedback of "tacit" knowledge between the various participants—and this tacit knowledge is in principle embedded only in human actors (see, e.g., Pavitt, 1991).

Institutions also differ in the two models. The linear model has fixed rules on which activities to be internalized and on what is to be exchanged between the three spheres. In the interactive model clearly there are rules, but many of them will vary from case to case, from collaboration project to collaboration project. In the case of knowledge transfer, one can describe the fundamental difference between the two models by saying that the linear model allows a type of (unilateral) transfer of codified knowledge, while the interactive model allows the transfer and feedback of both codified and tacit knowledge.

Defenders of academic independence usually refer to the traditional, linear model as a good guide, as it is in line with the idea of an independent academy that can maintain a critical attitude toward government authorities as well as the power of finance. However, if the discussion deals with economic growth, renewal and innovation, there are strong arguments that the interactive model creates much larger knowledge flows and more possible combinations of new knowledge and collaboration between actors, which should provide a greater potential for innovation.

The policy implications are clear. If government wants to support the establishment and strengthening of innovation, it can take action both to change

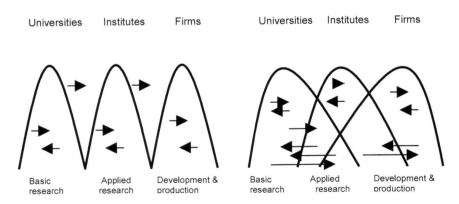

Figure 6.1 The linear (left) and the interactive (right) models for development of innovations.
Note: = Knowledge flows.
Source: Adapted from Arnold, Boekholt, Deiaco, et al., 2003; Westlund, Deiaco, & Johansson, 2005.

the institutions and to enhance learning processes. The first-mentioned can be achieved by removing obstacles to collaboration between academia, on the one hand, and business and other institutions with more applied goals, on the other, or by legislating that the university should have a third mission ("the collaboration mission" as it is called in Sweden). The latter can be achieved by allocating more resources to education and research, and to specific agencies and programs for collaboration and innovation.

The Spatial Dimension of Collaboration

According to economic theory, all interaction, exchange, and collaboration can be explained the interacting players getting more out of the interaction than it cost them. Expressed in economic terms, the inception, duration, extent, and frequency of the interaction/collaboration are determined by its costs and benefits.

The cost of interaction decreases with accessibility, which is dependent on distance, modes of transportation and communication, and other conditions of communication (e.g., language, education, administrative boundaries, time differences, etc.). This means that high accessibility in general contributes to greater collaboration and that collaboration between stakeholders within a region is greater than between them and extra-regional actors. This is the basic rationale for the existence of regional innovation systems.If an agent fails to find the optimal input from collaboration partners within the region, it has the option of finding a partner outside the region. Whether the optimal collaboration partner is outside the region varies greatly among different activities. For example, it can be assumed *ceteris paribus* that firms in mature industries with unskilled labour and low frequency of innovation are more likely to settle for local collaboration partners, while certain specialized innovative companies in the knowledge economy have only a few potential partners in the world. Another difference would be in the firms' markets. Small businesses operating in a local/regional market are in most cases likely to have a need for only local/regional partners. Major companies operating in major markets with fierce competition should however have a considerable need for collaboration partners globally to align products, marketing, and product development to national requirements and preferences. The same can be said about universities' collaboration with other actors in society. Regional universities collaborate mainly with actors in their own region, while larger ones collaborate nationally and internationally as well as regionally. One factor that to some extent limits the collaboration with actors outside the region is the increased need for frequent contact "face to face" in innovative activities.

In addition to the characteristics of the activity in question, the combination of intra-regional and extra-regional cooperation partners is determined by the characteristics of the region, such as population, economic size, the size of the industry/cluster/innovation system, and how well that system works. The latter is strongly linked to characteristics of the system, such as its governance and its social capital.

All the abovementioned factors mean that space's impact on innovation systems varies widely. Except for transportation and communication costs, the factors that operate in a spatially agglomerative direction can in a narrow sense be described in terms of "tacitness," i.e., whether the knowledge of the innovation system is non-standardized and tied to individuals and groups who constantly are refining it,[5] and the positive elements of regional social capital as a common "spirit" and tradition of cooperation. On the one hand, there may be a healthy competition mentality that encourages players and supports innovation. On the other hand, there may be factors that inhibit collaboration, such as organizational and administrative rivalry, legacy of distrust, personal conflicts, and other negative elements of social capital, which reduce collaboration and the region's innovation potential. In a broader sense, these factors can be expressed in terms of a region's "creativity potential" and the spatially bound social capital that make the region attractive for certain activities. There are many signs that regions with such properties, often industry-specific, tend to establish global systems of "gateways" with "knowledge pipelines" between them (see, e.g., Andersson & Andersson, 2000, and Wolfe, 2005). A reasonable interpretation of this phenomenon is that one cannot and should not investigate the regional innovation system in isolation, but must also include its external links in the analysis.

From a policy point of view, one often important issue is whether the innovation system is "improving" over time. The simple answer would be that more collaborations, more actors involved, more forms and types of exchanges, increased spatial coverage, higher frequency and intensity of collaboration, etc., always are positive signs of innovation, its quality, and innovation capacity. In reality, however, the situation is not so simple. Innovations are, with Schumpeter's words, "new combinations of factors of production." Each step in a process of innovation—from idea to finished product to market—depends on several factors: skilled labor, finance, knowledge, information, feedback, alternative ideas, etc. In today's complex, global economy, no single player has all these factors under the same roof; rather it is dependent on collaboration with others.

Both from the perspective of the individual actor and the innovation system, collaboration can be considered as something that, with diminishing marginal returns, up to a certain level promotes innovation. A possible hypothesis is that collaboration's importance for the innovation process can be described as an inverted U-curve, where the benefits of collaboration will increase up to a certain point and then decline. Also, it can be the case that "too much" collaboration in some dimensions and some areas—perhaps long-established ones—inhibits the necessary collaboration in other, perhaps newer areas. The collaboration in one area can make collaboration in other areas unnecessary or ineffective. One must also be aware that external financing from government programs can increase collaboration temporarily, but collaboration may decrease or even cease when the participating actors have to finance it themselves. Government programs in support of innovation are as a rule time restricted and if the collaborating actors cannot achieve sufficient benefits within the project time, they might not be interested in more resources in the project.

COLLABORATION TO FULFIL THE INNOVATION SYSTEM'S FUNCTIONS

Although we can imagine systems with "too much" collaboration in some (probably old) areas and too little in other (probably new) areas, and can assume that collaboration with the help of external resources temporarily can exceed what the actor's own resources, short-term needs, and ability would call for, normally there are probably good arguments that more collaborations with more actors are signs of innovation systems are improving. The reasons are several. First, it is reasonable to assume that the collaborating actors have considered pros and cons in putting resources into collaboration and have determined that the current collaboration is the best use of resources. Second, in the research literature there is a relatively wide consensus that innovation systems in the global knowledge economy involve more players and demand more and faster information exchange. More, especially new collaborations and more new actors and increased collaboration frequency and intensity could thus be signs that innovation systems are evolving with the global knowledge economy.[6]

The expected effect of increased collaboration in an innovation system ultimately should be sustained economic growth that can be measured by the industry's/region's contribution to GDP or GRP. Other expected, quantifiable effects are more innovations (often measured by patents), more firms with more employees, more research institutes with more staff, etc.

However, these clearly measurable effects of collaboration usually do not occur directly, but require some intermediary link of a qualitative nature. Important intermediary links are the ability to collaborate with actors one has not previously worked with and the ability to identify and solve problems. These abilities depend on various aspects of the social capital that dominate the innovation system and its subsystems. The concept of social capital, in the spirit of Putnam (1993), has often been described as something entirely good that can be found in the civil society and its organizations. However, a number of researchers have pointed out that certain social norms and networks may very well have adverse effects on both growth and democracy (Portes, 1998; Florida, 2002). It has also been pointed out that social capital does not exist only in the civil society but also in industry and other organizations, such as public sector bodies—and in innovation systems (Westlund, 2006).

In other words, the prevailing social capital affects the ability to collaborate in an innovation system. As in virtually all organizations, it is the leadership's norms, values, and networks, as well as its social competence (defined as the ability to use the social capital one has access to) that indicate how the ability to collaborate, identify problems, and solve problems develops.

COLLABORATION AND ITS EFFECTS

Marshall's (1890/1920) note on industrial districts and the "in the air" specialized knowledge in many ways did not bear fruit until about a hundred years later, when the concepts of clusters and innovation were coined by the various schools as embodiments of (partly) similar phenomena as those Marshall had observed. Although there are differences between the theories of industrial districts, clusters, triple helix, and innovation and the various theories also highlight other important variables, there is no doubt the theories emphasize various forms of collaboration between different actors as the core of success.

Simply put, it can be seen that the basic assumptions of these theories are confirmed every day. Firms co-operate with one another and with other actors in many ways (see Table 6.1. below), while they simultaneously compete in markets. Business and related activities concentrate in spatially distinct clusters. The main argument that collaboration is having positive effects for the collaborating actors is simply that they know what is best for themselves: if collaboration did not result in the expected net return, they would not keep on with it.

However, it is not reasonable to expect all types of collaboration always to have a positive economic impact. What kind of collaboration has the most positive impacts depends on a number of different characteristics of the innovation system and its actors. It is also possible that different types of collaboration substitute for each other, so that an increase in one type of collaboration causes a decrease in others.

What kind of research on the relationship between collaboration and innovation do we find? Roughly speaking one can distinguish two main approaches: a business administration dominated approach based on case studies of companies, networks, or specific cooperation projects at the regional or national level; an econometric approach based on large quantities of strata-sampled register data and survey data. The first approach has dominated as measured by the number of studies, and among other things it has given rise to the theories of innovation systems and related theories and numerous case studies. The latter approach has been more limited and practitioners have mainly devoted themselves to test these theories at the national level or in regional comparisons. Eurostat's *Community Innovation Survey* (CIS), implemented in all EU member states plus Norway and Iceland, is a body of data often used in these studies.

Each of the two approaches has its pros and cons. The advantage of case studies is that one can "go deep" and obtain quantitative and qualitative data one would not otherwise have access to and use those data to study possible correlations in detail. Case studies can also be used to generate hypotheses. The disadvantage of case studies is that they generally only provide high quality information on the studied cases, and often it is unclear how representative these cases are for the industry, region, or nation.

The advantage of the econometric approach is its use of large volumes of data and established statistical methods. The results will therefore be generalizable to a much greater extent. The disadvantages are that it is limited to available data, with the possible defects in them, and the approach does not provide any opportunity to "dig deeper."

Neither approach provides any unambiguous results. A general conclusion of the case study approach seems to be that in principle some form(s) of collaboration always have a positive association with innovation or other growth indicators, but the forms of cooperation vary. The results are, in other words, context-dependent.

NETWORKS, STRUCTURAL HOLES AND CRITICAL LINKS

Each actor in an innovation system takes part in different networks with varying aims and structures. The networks' aims are focused on the needs of the stakeholders participating in them and thus can be focused on supply, sales, marketing, R & D, knowledge exchange, management, administration, etc. Within the innovation systems there are likely networks with primarily social objectives, but most networks are naturally focused on the actors' core businesses. However, that does not mean the networks for marketing, R&D, knowledge transfer, etc., have no social aspects. Human actors maintain network connections, so the social relationships are often important if networks are to function optimally and transfer information and solve problems quickly.

Empirical studies of the Silicon Valley and other regional innovation systems have shown they actually are composed of different groups or sub-systems, more or less isolated from one another (Castilla et al., 2000). From a network perspective, therefore, a regional innovation system can be considered in at least two different ways. The system can either be seen as a spatial agglomeration of smaller networks, connected by some links between specific actors in the smaller constituent networks, or as a large, cohesive network with a number of subnets between which a varying number of links maintain connections.

Burt (1992) has coined the term "structural holes" for the gaps that exist between non-interconnected, isolated networks. Bridging the structural holes breaks the networks' isolation and contributes to new dynamics. A related concept is "critical links," which connect hitherto non-interconnected networks to a larger network and convert the previously isolated networks to sub-networks (Anderson & Strömqvist, 1988).

A regional innovation system's division into various sub-networks is in many ways appropriate and perhaps necessary. Most actors have contact with partners they think they need to deal with and they have no incentives to create and maintain a larger network than necessary. The presence of subnets can of course also be due to competition between different blocks of actors—something that can both stimulate and drive innovation but that also can lead to deficiencies in the dissemination of information and inefficient use of resources. Competition can be based on the struggle for the same inputs or markets, but it can also be based on personal and institutionalized antagonism that is being "learned" by the system. Although the existence of subnets within an innovation system therefore may have rational explanations and be an effective way to use resources, it is likely that actors' sense of belonging to different subnets also means there are fewer possible "combinations of production factors," implying that the actors' potential for innovation decreases, and that the system's overall productivity and efficiency becomes suboptimal.

VARIOUS DIMENSIONS OF COLLABORATION AND EXCHANGE—A TYPOLOGY

Collaboration between companies, research bodies, governmental agencies, and other actors in an innovation system can take an almost infinite number of different forms. A typology of the different dimensions of collaboration and variables is shown in Table 6.1. The number of dimensions and the large number of variables give a picture of the complexities embodied in an innovation system—and thus also the methodological challenges of studying them.

In Table 6.1., variables without any or with only negligible connection to social capital are shown in italics. The other variables, the vast majority, are to a greater or lesser extent affected by social networks and the values and norms being distributed in them. Thus, social capital is important regarding the *objectives and aims* of collaboration, since well-functioning social links and common norms, such as trust, facilitate finding common objectives for the collaboration. Concerning *who* are the collaborators, social relations with other agents facilitate and reduce interaction costs. When it comes to the *form of collaboration*, the table shows examples of one-directional exchange where social relations play a negligible role. However, most forms of collaboration and exchange have social elements in which the quality of the relations between the collaborating actors affects the form and outcomes. The *type of collaboration* is mainly centred on exchange of various types of knowledge and thus is affected by social relations and values. The *spatial extension* of collaboration, as mentioned above, depends on costs of transportation and communication and the "tacitness" of knowledge, but also on qualities of the social capital of the networks.

Table 6.1 Various Dimensions of Collaboration in Innovation Systems with Examples of Variables. Variables without or with Negligible Connection to Social Capital in Italics

Dimension	Variables
Objective and aim of collaboration	Marketing, distribution, sales, purchases, financing, R&D, product development, staff training, etc.
Collaboration with who/whom?	Suppliers, customers, competitors, financers, complementary actors, consultants, "brokers"/intermediaries, R&D bodies, agencies, third sector organizations, etc.
Form of collaboration and exchange	· *Data banks, journals, etc.* · Informal and formal social networks · Informal and formal networks for interchange on issues of a scientific, technological, or managerial nature · *One-directional technology flows* *technology licenses purchase/sale* *sub-licensing* *second sourcing agreements* *assets purchase/sale* *purchase/sale of contract research for product/process development* *purchase/sale of contract research for clinical trials* · Customer-supplier relations purchase/sale of material, part, product purchase/sale of manufacturing purchase/sale of distribution purchase/sale of marketing · Agreements on technology exchange cross-licensing technology sharing options mutual second sourcing · Joint R&D agreements co-development co-production co-marketing joint venture · Direct investment minority holdings mergers and acquisitions
Type of collaboration and exchange	Exchange of: · Technology-related resources and knowledge Personnel, initial technological opportunity, technological ideas, intellectual property, supplementing technology, products, parts, process technology, access to clinical trials, know-how of employees, etc. · Production-related resources and knowledge Personnel, equipment, facilities, etc.

Continued

Table 6.1 Continued

Dimension	Variables
	· Management-related resources and knowledge Personnel, methods, know-how of employees, etc. · Market-related resources and knowledge Personnel, market knowledge, knowledge of regulations, education of customers, market access, distribution channels, contracts, trade secrets, know-how of employees, etc. · Credibility-related resources and knowledge Legitimacy, brand name, social capital, know-how of employees, etc. · Financially related resources and knowledge Personnel, initial financing, financing for firm development, know-how of employees, etc.
Spatial extension	Distance between actors, intra-/interregional links, international links, etc.
Form of agreement	Alliance, contract, informal agreement, etc.
Frequency	Number of collaborations, contact frequency
Intensity	Number of contacts during a certain time
Time dimension	Duration, long/short term
Number of collaborating actors	From the single actor's or the network's/system's perspective
Actors' characteristics	*Type of industry, technology intensity, knowledge intensity, actor's size, regional/national/trans-national actors, etc.*
Type of network for collaboration	Vertical/top-down/forwards-backwards, horizontal.
Financing of collaboration	Share of owners' financing, bank loans, venture capital, governmental subsidies, etc.

Source: The table is an elaboration based on a number of contributions, among them Levy et al. (2006), Skov Kristensen (1999), and Pavitt (1984).

The *form of agreement* on collaboration depends on, among other things, the strength of actors' social relations. The same holds for the *frequency, intensity, and time dimension* of collaboration. The *number of collaborating actors* is very context dependent and thus depends on, among other things, social relations. The *characteristics of the collaborating actors* are one of the exceptions in Table 6.1., as in the main they are affected only indirectly by social capital features. The *type of network* for the collaboration is highly connected to social relations in the constituent networks, although it is more likely that the structure of the collaboration networks that shapes the social relations, rather than the reverse. Finally, the *financing* of collaboration is a dimension for which social capital can play an important role.

CONCLUDING REMARKS

With a focus on the collaboration processes of innovation systems, this chapter has provided an introductory analysis of the role of social capital for various dimensions of collaboration. The results show that social capital is important for the performance and outcomes of a large number of collaborations. A conclusion is that innovation policies should pay greater attention to the social capital of innovation systems.

However, there are several steps between introductory theoretical analysis and implementation of policies. A first step would be to develop empirical analyses of how actors' social networks and relations and values, norms and attitudes influence various dimensions of collaboration in innovation systems. There are of course already a large number of case studies in many of the dimensions listed in Table 6.1., even if many of them are not placed within the theoretical framework of social capital. Thus, there is a potential for meta-studies. However, the fact that existing case studies have not been coordinated means that the comparability of the results will vary. The lack of coordinated data collections thus sets a limit for how much new knowledge that can be expected from meta-studies on existing studies.

More advanced knowledge of the role of social capital in innovation systems requires coordinated collection of empirical indicators of actors' social capital, adapted to the various dimensions of collaboration. From a methodological point of view this means big challenges. One possibility would perhaps be to organize a European research program on data collection of collaboration in innovation systems.

A second step between theory and policy implementation would be to analyse policies' potential to influence social capital among actors of innovation systems, in the form of behaviour and attitudes to various dimensions of collaboration. This requires both theory development and empirical studies. Regarding the latter, also these analyses can start with meta-studies of existing studies and evaluations of innovation programs and projects. However, similar to the first step, obtainable studies are based on programs and projects that seldom are wholly comparable. Still, meta-studies of this problem would bring new, state-of-the-art knowledge on policies' possibilities, and generate hypotheses for further research.

In a third step, policies with the deliberate aim to influence the social capital of innovation systems can be developed, formed, implemented, and evaluated. The research agenda outlined here may seem very ambitious and hard to realize completely. It is of course necessary to take one step at a time. Starting with the meta-studies will probably give interesting results to go further with.

NOTES

1. This means that innovations that are not directly commercial, as, e.g., in innovations in administration, marketing and education are included in our definition.
2. Some theories, e.g. triple-helix, only include some of the actors. However, it has been suggested to extend these models to so called models of "Mode 3" or "quadruple-helix" (see Carayanis & Campbell, 2011) in which also civil society is included.
3. These applied theories are in their turn, in varying extension being based in institutional economic theory, evolutionary economic theory, endogenous growth theory and the new economic geography.
4. Rutten and Boekema (2004) even claim that if universities cannot adapt to society's changed demand for knowledge, society will marginalize them and allocate its resources for knowledge production to other actors.
5. Correspondingly, the importance of proximity decreases if the particular knowledge is available in codified form that easily can be transferred over long distances (see, e.g., Wolfe, 2005).
6. Intensifying of established collaborations can on the other hand be an indication of undesirable "lock-ins"; see, e.g., Grabher (1993).

REFERENCES

Andersson, Å.E., & Andersson, D.E. (eds.). (2000) *Gateways to the global economy*. Cheltenham: Edward Elgar.

Andersson, Å.E., & Strömquist, U. (1988) *K-Samhällets Framtid*. Stockholm: Prisma.

Arnold, E., Boekholt, P., Deiaco, E., et al. (2003) Research and innovation governance in eight countries—A meta-analysis of work funded by EZ (Netherlands) and RCN (Norway). Technopolis. *www.technopolis-group.com*. Accessed June 2005. .

Burt, R.S. (1992) *Structural holes: The social structure of competition*. Cambridge: Harvard University Press.

Carayannis, E.G., &Campbell, D.F.J. (2011) Mode 3 knowledge production in quadruple helix innovation systems: 21st-century democracy, innovation, and entrepreneurship for development. *Springer Briefs in Business*, 7.

Castilla, E.J., Hwang, H., Granovetter, E., & Granovetter, M. (2000) Social networks in Silicon Valley. In Lee C.-M., Miller, W.F., Gong Hancock, M., Rowen, H.S. (eds.), *The Silicon Valley edge: A habitat for innovation and entrepreneurship*. Stanford: Stanford University Press, 218–247.

Edquist, C. (1997) Systems of innovation approaches—Their emergence and characteristics. In Edquist, C. (1997) (ed.), *Systems of innovation: Technologies, institutions and organizations*. London: Pinter/Cassel.

Florida, R. (2002) *The rise of the creative class: And how it's transforming work, leisure, community and everyday life*. New York: Basic Books.

Freeman, C. (1987) *Technology policy and economic performance: Lessons from Japan*. London: Pinter.

Gibbons, M., Limoges, C., Nowotny, H., Schwartzman, S., Scott, P., & Trow, M. (1994) *The new production of knowledge: The dynamics of science and research in contemporary societies*. London: Sage.

Grabher, G. (1993) The weakness of strong ties: The lock-in of regional development in the Ruhr area. In G. Grabher (ed.), *The embedded firm: On the socioeconomics of industrial networks*. London: Routledge, pp. 255–277.

Grupp, H. (1998) *Foundations of the economics of innovation—Theory ,measurement and practice.* Cheltenham: Edward Elgar.
Levy, R., Roux, P., & Wolff, S. (2006) *A study of science-industry collaborative patterns in a large European university.* Paper presented at the DRUID summer conference, Copenhagen, Denmark, June 18–20, 2006.
Lundvall, B.-Å. (1988) Innovation as an interactive process: From user-producer interaction to the national system of innovation. In Gosi, G., Freeman, C., Nelson, R., Silverberg, G., & Soete, L. (eds.), *Technical change and economic theory.* London: Pinter.
Lundwall, B.-Å. (1992) (ed.). *National systems of innovation: Towards a theory of innovation and interactive learning.* London: Pinter Publishers.
Marshall, A. (1890/1920) *Principles of economics: An introductory volume,* 8th edition 1920. London: Macmillan.
Maskell, P. (2000) Social capital, innovation and competitiveness. In Baron S, Field, J,, &Schuller, T. (eds.), *Social capital: Critical perspectives.* Oxford: Oxford University Press.
McDaniel, S.E. (2006) *Where science, technology and innovation indicators hit the road and roadblocks.* Paper presented at the OECD Blue Sky II Forum "What Indicators for Science, Technology and Innovation Policies in the 21st Century," September 25–27, 2006, Ottawa.
Nowotny, H., Scott, P., & Gibbons, M. (2001) *Re-thinking science: Knowledge and the public in an age of uncertainty.* Cambridge: Polity Press.
Pavitt, K. (1984) Sectoral patterns of technical change: Towards a taxonomy and a theory. *Research Policy,* 13, 343–373.
Pavitt, K. (1991) What Makes basic research economically useful? *Research Policy,* 10, 109–119.
Portes, A. (1998) Social capital: Its origins and applications in modern sociology. *Annual Review of Sociology,* 24, 1–24.
Putnam, R.D. (1993) Making democracy work. Civic traditions in modern Italy. Princeton: Princeton University Press.
Rutten, R., & Boekema, F. (2004) *Universities and the knowledge-based economy: The (ir)relevance of universities in a networked, regionalized world economy.* Paper presented at the Western Regional Science Association's 43rd annual meeting. Maui, Hawaii, February 25–29, 2004.
Skov Kristensen, F. (1999) *Towards a taxonomy and theory of the interdependence between learning regimes and sectoral patterns of innovation and competition: An empirical analysis of an elaborated Pavitt taxonomy applying Danish data.* Paper presented at the DRUID Winter conference, Seeland, Denmark, January 7–9, 1999.
Westlund, H. (2006) *Social capital in the knowledge economy: Theory and empirics.* Berlin. Heidelberg. New York: Springer.
Westlund, H., Deiaco, E., &Johansson, M. (2005) *Utvärdering av Delegationen för regional samverkan om högre utbildning* (Evaluation of the special committee for regional cooperation on higher education). Östersund: ITPS A2005, 14.
Wolfe, D.A. (2005) *The role of universities in regional development and cluster formation.* http://www.utoronto.ca/progris/pdf_files/Higher_Education_UofT-Press.pdf. Accessed April 2007.

7 Learning in Regional Networks
The Role of Social Capital

Roel Rutten and Dessy Irawati

This chapter deals with the question, why do some regions perform better in the global economy than others? Given Porter's (1990) adage that not nations (or regions) compete but companies, and given that successful companies are often embedded in strong geographically concentrated networks of companies (Dupuy & Torre, 2006, Scott & Storper, 2003), the question is more accurately rephrased as: Why do some regional networks perform better than others? To answer that question, this chapter follows two related but distinct lines of inquiry. In the first place, the characteristics of regional networks are important for understanding their success or failure in the global economy. Second, also the characteristics of the region wherein a network is embedded must be considered. Both lines of inquiry have been well-researched in the economic and economic-geographic literatures in recent decades; however, this chapter argues that one crucial element has remained underdeveloped: social capital. Social capital can be found in the literature under a variety of names, such as bonding and bridging social capital (Burt, 2005), trust and the social depth of networks (Morgan, 2004; Storper & Venables, 2004) and norms and values (Florida, 2002; Inglehart & Baker, 2000), and typically pertains to two levels of analysis. These two levels are reflected in the above-mentioned lines of inquiry, that is, the level of (networks of) individuals and the level of the regions in which these networks are embedded. Social capital is crucial to understanding the performance of networks in the global competition for the following reasons. Competitiveness in today's knowledge economy is fundamentally driven by innovation, learning, and knowledge creation. Therefore, the degree to which a regional network is successful in creating new knowledge and its application in innovations is of crucial importance for the competitiveness of the network and the companies within it (Best, 1990; Oerlemans et al., 2007). The degree in which a region offers a favourable social and institutional environment for learning and innovation contributes critically to this process (Dupuy & Torre, 2006; Scott & Storper, 2003). Since learning is fundamentally a matter of interaction between individuals (from different companies) social capital is of crucial importance but has not been given this place in the literature so far (Rutten et al., 2010). This chapter makes a case for the role of social capital by identifying how it plays a role with respect to learning in regional networks.

This chapter is structured in five sections. The first section explains the role of social capital with regard to learning in networks. The second section presents a short case study of the Western Java automotive industry. This case study is relevant as social capital plays a different role in this industry than it does in similar networks in regions in the developed world. The third section takes a closer look at the spatial dimension of networks and social capital. Recognizing that social capital is also found at the level of the region, the fourth section elaborates how these two levels—(networks of) individuals and regions—can be conceptually connected with regard to the role of social capital in relation to learning. This section argues that the regional level is not a good starting point for this discussion. The final section presents a summary and conclusion of the arguments brought forward in this chapter and suggests that an 'individual turn' is needed in economic geography to explain the role of social capital with regard to learning in networks.

NETWORKS, LEARNING, AND SOCIAL CAPITAL

The nature of the global economy as a network economy has, among others, led to the rise of geographically concentrated networks playing an important role. Such networks provide a constructive and efficient opportunity for discussion among related companies, their suppliers, government, and other institutions (Breschi & Malerba, 2005; Nooteboom, 2006). Because of externalities, public and private investments in the network and its regional environment benefit many firms and can make substantial contributions to a region's economy (Hudson, 1999). The advantages that accrue to companies by being part of a regional network have been summarized in the mainstream economic (geography) literature as follows (cf. Irawati, 2009):

> Access to specialized inputs and employees. Regional networks can provide superior or lower cost access to specialized inputs such as components, machinery, business services, and personnel compared to vertical integration, formal alliances, or importing inputs from distant locations (Best, 1990; Porter, 1990; Teece, 2000).

> Access to information. Proximity, supply, and technological linkages, and the existence of repeated personal relationships and community ties fostering trust, facilitate the information flow within a regional network (Dupuy & Torre, 2006; Uzzi, 1997).

> Complementarities. Regional networks increase productivity not only through the acquisition and assembly of input but also through facilitating

complementarities between the activities of network members (Best, 1990; Breschi & Malerba, 2005; Lazerson & Lorenzoni, 1999).

Access to institution and public good. Firms within a regional network can access specialised infrastructure, or advice, from experts in local institutions at very low cost.

Incentives and performance measurement. Regional networks improve the incentives within companies. 1) Competitive pressure, when the pride and desire to perform well in a local community motivate firms to compete with one another. 2) Regional networks also make it easier to measure the performance of in-house activities because there are local firms that perform similar functions.

However, a key aspect of networks, their social capital, is largely missing from this argument. Social capital is a characteristic of the relations between individuals in networks and the social and economic benefits that result from them (Westlund & Adam, 2010). Social capital performs two key functions, that of a glue and that of a lubricant. As glue, social capital binds people together in relations of mutual dependency. As a lubricant, it facilitates social interaction in networks through shared norms and values (Rutten & Gelissen, 2010). In general, social capital is argued to be a good thing; however, some forms of social capital can have highly detrimental effects on networks and the individuals within them. As learning is a matter of social interaction (Amin & Cohendet, 2004), it is unsurprising that a clear relation has been identified between social capital and learning in networks (Hauser et al., 2007; Rutten et al., 2010). The function of social capital in human work relations had first been described by French sociologist Emile Durkheim in his *The Division of Labour in Society*. He argued that "It is impossible for man to . . . be in regular contact with one another without their acquiring some feel for the group . . . without their becoming attached to it" (Durkheim, 1893/1997, xliii).

A more elaborate of explanation of social capital comes from the discussion of bonding and bridging social capital (Burt, 2005; Field, 2003). Bonding social capital refers to a situation where there are strong linkages between all or most of the individuals within the network; this produces higher levels of trust and shared norms and values which, in turn, are helpful for learning and innovation. In networks with a high level of bonding social capital companies are likely to find a substantial number of trusted partners for learning and innovation, which may result in a better economic performance of these networks. Higher levels of bridging social capital also have a positive effect on learning and innovation. Bridging social capital refers to a situation where a limited number of individuals in a network have strong relations to individuals in other networks. In other words, these individuals act as bridges between two (or more) networks, which are extremely relevant to transfer new knowledge and ideas into the network. Networks must have linkages to other networks to access knowledge not available inside the network. But too many such links may be detrimental as they can lead to knowledge overload. In sum, networks that have both high levels of bonding and bridging social capital are the most conducive learning environments.

Many (case) studies have appeared that demonstrate how social capital contributes to learning and innovation in networks. In a case study of the New York garment industry Uzzi (1997), for example, showed that trustful relations make for richer and thicker flows of knowledge between network partners which facilitates innovation. Successful entrepreneurs in his study had networks that were rich in both bonding and bridging social capital. Larson (1992) too showed the importance of the social capital (without actually using this term) for transactions in explaining control and coordination in exchange relations. Statistical research by McFadyen et al. (2009) also demonstrated that learning in networks relies in part on network structure, tie strength, and network density. What these and other studies have in common is that they show beneficial effects of social capital in networks. The next section discusses the automotive industry in Western Java, Indonesia, and shows how social capital—or a lack thereof—can hamper learning in networks.

THE WESTERN JAVA AUTOMOTIVE INDUSTRY

In recent decades the automotive industry in Western Java has successfully followed a knowledge-based development path and has become a key driver of the regional economy (Irawati, 2009; Irawati & Charles, 2010). Following deliberate attempts to create regional networks of related companies to further the development of these companies, this regional network is often referred to as the automotive cluster. In addition to inter-firm linkages, the cluster also has strong linkages to several leading public and private regional knowledge centres as well as intermediary organizations and government agencies (cf. Morgan & Nauwelaers, 2003; Lagendijk & Charles, 1999).

Characterising the weight of the automotive cluster in the Western Java region is somewhat problematic because of limited availability of (regional) statistics. However, several official documents suggest that (nearly) all of Indonesia's automotive industry is clustered in the West Java region (MOSR, 2002; MOTIRI, 2005). Depending on the source this industry accounts for 64,000[1] to 90,500 jobs.[2] Although the direct impact of the automotive industry on the regional economy seems limited given its relatively modest size, the cluster plays an important role in the ASEAN automotive market, where it has a market share of more than 25%. Moreover, the industry is an important exporter to South America and the Middle East.[3] Becoming a notable international exporter from being a mere provider of cheap production capacity in the 1970s and 1980s is one of the cluster's greatest achievements and is a good indicator that this cluster's relevance for the regional economy may be more substantial than its employment share suggests (Irawati, 2009; Irawati & Charles, 2010). An important characteristic of the cluster is the dominant position of Japanese car makers. (Nine out of every 10 cars sold in Indonesia in 2009 was a Japanese car.[4]) In particular, the networks of Toyota and Honda are the pivot around which the cluster turns (Irawati & Charles, 2010).

The Japanese domination of the Western Java automotive cluster results from large FDI of Japanese car makers and the cluster's dependence on its technological and managerial knowledge. From the 1970s onward, Japanese car makers, Toyota and Honda in particular, established 'transplants' in the Western Java region: that is, companies under Japanese ownership and management employing production technology transferred from their Japanese parent companies. The transplants served as 'workbenches,' to achieve cost benefits for their parents (Irawati, 2009; Ozawa, 2005). In the following stages the transplants started to outsource production to local subsidiaries and subcontractors, thus establishing separate regional production networks around individual Japanese car makers. In order to bring them up to Japanese standards the local Indonesian firms received FDI in the form of managerial, organizational, and technological knowledge transfer from the Japanese car makers (Irawati & Charles, 2010). The knowledge transfers came with long-term commitments that effectively incorporate the local firms into the production networks of 'their' respective car makers. Toyota and Honda and to a lesser degree other Japanese car makers have thus "established a partially internalized market for intermediate products" (Gerlach & Lincoln, 1992, 493) which results in stable, long-term network relations (Hatch & Yamaura, 1996; Ozawa, 2005). This strategy allowed the Japanese car makers to spread some of the costs and risks of doing business in the highly competitive automotive market while offering stable long-term network relations to local firms that provide much more certainty than market relations. The FDI of Japanese car makers to upgrade the firms in their respective regional production networks probably benefited the former more than the latter (Gerlach & Lincoln, 1992, Gwyne, 1990, Hatch & Yamamura, 1996; Ozawa, 2005). The resulting dependency of the Western Java automotive cluster on in particular Toyota and Honda can hardly be underestimated (Irawati, 2009; Irawati & Charles, 2010).

Local subsidiaries and subcontractors are committed to using precisely calibrated tools and dies for the production of the parts and subassemblies outsourced to them. To operate such sophisticated equipment, the Indonesian firms have to make substantial investments in highly specific job training programmes for their employees (Hatch & Yamamura, 1996; Irawati, 2009; Pries & Schweer, 2004). The resulting asset specificity and sunk costs make it increasingly unattractive for the Indonesian firms to work for other MNEs. Moreover, such a move is likely to upset the Japanese 'parent,' which brings the real danger of the relationship being terminated, resulting in a dramatic reduction of the value of the investments and the assets of local Indonesian firms (Chen, 1996; Doner, 1991; Hatch & Yamamura, 1996; Honda, 2004; Toyota, 2007). The aspect of loyalty needs to be considered as a part of the network relations. Loyalty plays a different role in an Asian context than it does in the West (Gerlach & Lincoln, 1992) and it enables the Japanese car makers to exercise control over the Indonesian firms in their respective production networks in a far more subtle and probably more effective way than may be achieved by traditional control mechanism such as organizational, managerial, technological, and legal control. The resulting coordinated deployment of resources in the various production networks, in particular managerial and technological knowledge transfer, allows the Japanese firms to capture more firmly the gains from their resources. Knowledge spillovers are largely intentional and internal to the various production networks. Consequently, knowledge spillovers between production networks are far less frequent than is the case in similar regional clusters of American and European car makers (Busser & Sadoi, 2004).

In sum, the automotive cluster contributes to transforming the low-wage, labour-intensive, inward-looking economy of the Western Java region into a higher-wage, technology-intensive, export-oriented economy. However, it comes at a price. The developmental path of separate regional production networks has a negative impact on the development of linkages between these networks, which stifles local initiatives and the potential for endogenous growth (Irawati & Charles, 2010). One of the crucial differences between the Western Java automotive industry on the one hand and similar regionally clustered networks in the developed world is that the Java network has been purposely created while such networks have usually organically grown in developed economies. This works out negatively for the Western Java automotive industry for the following reason. The economic geography literature emphasises the value of (partially) overlapping social and economic networks in a region (Lorenzen, 2007; Moulaert & Sekia, 2003; Morgan, 1997; Storper, 1993). This is because social capital (e.g., norms, values and customs, societal institutions) may facilitate interactive learning in economic (exchange) relations (Lazerson & Lorenzoni, 1999, Uzzi, 1997). However, as the case of the Western Java automotive industry shows, social and economic networks may not overlap when economic networks were purposely created in a top-down fashion rather than organically grown. The social institutions governing economic relations in the Western Java automotive cluster are to a large degree handed down from Japan rather than picked up from the local context. This Japanese social capital favours hierarchical relations between companies, loyalty, and obedience while discouraging bottom-up initiatives. Moreover, social capital that plays a role in society is overruled by company codes on the workplace. Social capital built in society, therefore, only marginally permeates into economic life in the Western Java case.

Although this situation is not unique to developing countries—in fact, a study by Rugraff (2010) shows a similar network configuration and accompanying dominance of network social capital over societal social capital in the Czech motor vehicle industry—the problem may be more compelling in developing countries as their industrial networks are more often designed rather than having emerged from (pre-industrial) underlying social and economic relations. In any case, the example of the Western Java automotive industry shows a regional network wherein economic exchange relations are to a very substantial degree disconnected from underlying social (societal) networks. In this the Western Java automotive cluster is different from examples of successful clusters in Europe and North America as discussed by, for example, Lagendijk and Charles (1999), Rutten and Oerlemans (2009), and Storper (1993).

In sum, the case of the Western Java automotive industry demonstrates the role of social capital in network relations. It also suggests that there is a link between the social capital in the network and the social capital that is connected to the region in which the network is embedded, that is, societal social capital. This spatial dimension of social capital is the subject of the next section.

THE SPATIAL DIMENSION

Talking about regional networks leads to the question of the importance of the spatial dimension. The answer to this question follows from two lines of inquiry. In the first place the role of the regional context must be considered, that is, the characteristics of the region in which the network is located. Second, it is important to consider the relevance of spatial proximity between network partners with regard to learning. The question regarding the relevance of the regional context is easily answered: It matters substantially (Best, 1990, Morgan, 2004). The regional business environment plays a crucial role in the economic performance of regional networks as some business environments are more conducive to learning and innovation than others. The literature in this field suggests four key characteristics of regional business environment that are important with regard to learning and innovation (Scott & Storper, 2003, Porter, 1990, Teece, 2000):

1. The provision of a physical and digital infrastructure and the tax and legal systems provide the bare basics for companies to operate in the 21st century.
2. The education levels of the regional workforce and the presence of public and private knowledge centres are of critical importance in today's knowledge economy.
3. High-quality demand of local customers (both companies and consumers) forces companies to innovate in order to meet that demand.

4. Rivalry among regional firms also encourages companies to be innovative in order to stay abreast of their competitors.

The economic geography literature offers compelling evidence that regional networks in favourable business environments perform (much) better in terms of learning and innovation than networks embedded in a less favourable business environment. In fact, much regional economic development policy is aimed at putting in place or improving the above characteristics. However, a local business environment offering sufficient but not great conditions may actually be equally helpful to the development of companies as so-called selective disadvantage trigger companies to innovate. Vice versa, a very comfortable business environment may render companies complacent (Porter, 1990; Teece, 2000). The key argument, thus, is that there is a relation between the level of sophistication of the local business environment, on the one hand, and the performance of regional networks, on the other hand (Nooteboom, 2006).

Regarding the second issue, the relevance of spatial proximity in relation to learning, the answer is much harder. The debate on this issue falls within two equally problematic extremes. One extreme argues for the 'geography of knowledge,' which claims that tacit knowledge can only be effectively communicated in face-to-face interactions which for reasons of efficiency require that partners are co-located. This argument has been refuted because, on the one hand, temporary proximity allows for effective face-to-face communication just as well as permanent proximity (Grabher, 2004). On the other hand, knowledge is no longer regarded as being tacit (or codified) but as context depended (Amin & Cohendet, 2004; Morgan, 2004). To the extent that partners share this (social) context, they can communicate even highly complex knowledge using digital media. Transferring of this knowledge to individuals outside the (social) context on the other hand requires intensive face-to-face communication for which temporary proximity suffices. The other extreme position concerns the argument of the 'death of distance,' which claims that digital means of communication allow knowledge to be exchanged regardless of distance (Amin & Cohendet, 2004). However, this argument conflates reach of communication and social depth (Morgan, 2004). Local norms, values, and customs shape the way people interact with one another. Being unfamiliar with local norms, values, and customs hampers the ability of outsiders to exchange knowledge with people within a regional network (Lorenzen, 2008; Morgan, 2004). Although knowledge exchange may benefit from shared professional norms and values, shared local social capital (societal social capital) is found to have an effect on economic development (Inglehart & Baker, 2000; Rutten & Gelissen, 2010) and also on knowledge exchange (Dupuy & Torre, 2008; Hauser et al., 2007; Storper & Venables, 2004).

Criticizing the two extreme positions makes clear the role of spatial proximity between network partners with regard to knowledge and learning.

This role is closely related to social capital in the form of norms, values, and customs as learning in networks is smoother when shared among network members. However, it would be a mistake to see norms and values (social capital) as characteristics of a region. Even though norms and values differ from one place to another they are characteristics of networks first of all. Norms and values are connected to places because the individuals making up the networks are largely spatially sticky. Most human beings are connected to the place they live, work, and have their friends and relatives; that is, to the place they call home. Consequently, social interaction (inclusive learning) is also spatially sticky. However, human networks frequently cut across different spatial scales; this makes it difficult to connect norms and values to places. Nonetheless, certain norms and values are more conducive to learning and innovation than others. Even though norms and values are characteristics of networks rather than regions, given that networks consist of spatially sticky individuals certain norms and values may be more prominent in some regions than in others. In fact, research shows a relation between the presence of certain types of norms and values and innovation and economic development of regions (Huntington & Harrison, 2000; Inglehart & Baker, 2000; Rutten & Gelissen, 2010).

The effect of norms and values on regional economic development cannot be isolated from more conventional variables, such as investments in research and development (R&D), human capital, and urbanization. Regional R&D investments and patent data have proved to be solid indicators of regional economic development in a large number of studies. This underlines that in today's economy economic development is fueled by learning and innovation (Morgan, 1997). Regional human capital, often measured as the percentage of the regional workforce with higher education, is an obvious complement to the above argument as learning and innovation are predominantly carried out in occupations that require higher levels of education. In many studies, therefore, regional human capital shows a strong correlation with regional economic development. Urbanisation is another factor that is directly related to regional economic development. Most economic activities take place in cities. Nonetheless, several empirical studies strongly suggest that norms and values matter. Rutten and Gelissen (2010), for example, found that regional economic development, measured as gross domestic product (GDP) per capita in purchasing parties, is largely explained by innovation, human capital, and urbanisation. Norms and values only have a limited direct effect on GDP. However, norms and values do have a considerable effect on innovation. This is a very significant finding as it suggests that regions harbouring norms and values that favour innovation are likely to be more economically developed than other regions.

The findings of Rutten and Gelissen (2010) corroborate those of other research (e.g., Beugelsdijk & Van Schaik, 2005; Florida, 2002; Inglehart & Baker, 2000). These studies suggest that norms and values reflecting a cosmopolitan attitude, self-expression, a move away from traditional and religious values, and a readiness to embrace new developments are the kinds of norms and values that encourage innovation. Tolerance for non-conformist behaviour and tolerance for socio-, cultural, ethnic diversity represents another set of norms and values that are strongly related to innovation. These values open people's minds for new ideas and encourage an atmosphere of creativity. Regions where such norms and values are prevalent contribute to innovation in two additional ways. In the first place they attract so-called knowledge workers, that is, the kind of workers who are engaged with learning and innovation as part of their jobs. Second, these norms and values and the socio-cultural diversity accompanying them represent a richer, more diverse and more challenging market which requires innovation to meet it (Florida, 2002; Rutten & Gelissen, 2010). Finally, participation in various social networks—such as professional, leisure, religious, political, and voluntary organizations—is related to both innovation and GDP. Although social networks do not represent norms and values as such, they are nonetheless an important example of social capital. Social networks may reduce transaction costs in economic life because of the trust and reputation effects that such networks generate. Social networks may also encourage the exchange of knowledge and ideas and thus contribute to creativity and innovation (Uzzi, 1997).

In short, the issue of the spatial dimension addressed in this section ultimately concerns the spatial dimension of social capital. According to Rutten et al. (2010), social capital is developed and maintained in social relations, that is, networks, "and to the extent that social relations are spatially sticky, so are norms and values . . . [i.e., social capital]" (p. 869). Accordingly, social capital can be studied on the level of networks as well as on the level of locations. The argument is that network relations will benefit (or suffer) from the prevailing social capital, in particular norms, values, and customs, of the location they are embedded in (cf. Morgan, 2004; Storper & Venables, 2004). The obvious place in the economic geography literature where regional social capital in relation to learning and innovation may be discussed is in the literature on Territorial Innovation Models (TIMs), such as the learning region, regional innovation systems, innovative milieu, and industrial districts. The TIM literature aimed to open the black box on the relation between space and learning by marrying two literatures, the networks and innovation literature and the literature on regional development. However, TIMs suffer from two weaknesses; in the first place they are subject to strong conceptual criticism and, second, social capital has never featured very prominently in them. The next section discusses the key criticism against the TIM literature and explains how a discussion of social capital may augment it.

TERRITORIAL INNOVATION MODELS AND SOCIAL CAPITAL

The TIM literature offers a way of thinking about the relation between innovation and regional development. Different TIMs highlight different aspects of the same idea, namely, that "regional economic success is heavily based upon territorially defined assets, derived from 'unique,' often tacit, knowledge and competitive assets, and stresses the importance of spatial proximity in collective learning processes" (Hudson, 1999, 64). TIMs stress the importance of informal rather than contractual relations between regional actors and thus place "the emphasis on shared values, meanings and understandings, specifically territorially embedded, and tacit knowledge and the institutional structures through which it is produced" (Hudson, 1999, 64). The overlap of social and economic networks greatly contributes to (social) institutions permeating to economic life and facilitating economic interaction between regional agents (Lorenzen, 2007; Morgan, 1997; Storper, 1993; Storper & Venables, 2004; Uzzi, 1997). Morgan (1997) explicitly mentions the role of regional institutions as a characteristic of the learning region, but other TIMs too recognize their relevance (Oerlemans et al., 2007). In any case, the differences between the TIMs are gradual and all of them suffer from considerable conceptual and empirical problems (Oerlemans et al., 2007). Given the lack of robust definitions of key concepts, such as 'innovation milieu' or 'innovation system,' a seemingly straightforward question as "how do we know a learning region or regional innovation system when we see one?" has proved very difficult to answer. Consequently, rigorous empirical research on TIMs is difficult to conduct. In fact, Moulaert and Sekia are of the opinion that their conceptual flexibility pushes all of the TIMs "across the border of coherent theory building" (2003, 289). Nonetheless, TIMs do shed light on the relation between space and learning and in that sense they are a welcome addition to the mainstream economic literature, which is largely a-spatial. The central premise of the TIM literature may be summarized as follows: Regional development follows from interactive learning in networks of regional agents capitalizing on territorially defined knowledge assets, such as formal and informal institutions and region-specific technological competences (Moulaert & Sekia, 2003; Oerlemans et al., 2007).

Although this position is open to the influence of social capital on the relation between space and learning, social capital is not systematically addressed in the TIM literature (Lorenzen, 2007). The key reason for this is that social capital is not in the first place a spatial concept but a relational concept. The spatial dimension of social capital may be thought of as being twofold. In the first place, social capital is connected to relations among individuals and to only to the extent that these individuals are 'spatially sticky' so is social capital. Second, since people living in the same place will entertain all sorts of social and economic interactions, they will develop a set of norms, values and customs (i.e., social capital) to guide their interactions. Social capital may give a place a particular appeal that in turn can facilitate learning, as was discussed in the previous section. The TIM literature recognizes the importance of social capital but this literature is not informed by a considered conceptualization of social capital and its spatial dimension(s). The position of the TIM literature that social capital is in some way connected to space is thus correct, but this position fails to recognize the fact that social capital is a relational concept and not a spatial concept. Since the majority of empirical TIM studies have been conducted in the industrial regions of Europe and North America, the TIM literature may have been blind to its problematic dealing with social capital. Social and economic relations in these regions overlap to a large degree (e.g., Morgan & Nauwelaers, 2003; Storper & Venables, 2004; Uzzi, 1997), which may give the impression that there really is something such as regional or local social capital. When trying to connect this 'regional' social capital to regional learning, the conceptual problems get even worse.

Taking the region as level of analysis, the TIM literature assumes a relation between what actors in that region do (learning) and the learning outcomes (innovations) that the region produces. It thus creates a black box by conflating the two levels of analysis introduced earlier, (networks of) actors and regions. Learning outcomes that materialize in a particular region may be the result of processes that originate in different places altogether. Learning processes will rarely, if ever, be strictly regional. In fact, innovative regions may be seen as nodes in global knowledge creation networks (Lorenzen, 2008; Storper & Venables, 2004). The reason for this is twofold. First, learning (or knowledge creation) happens in networks that may or may not have a regional dimension but even if they do learning is still a characteristic of these networks not of regions. The TIMs' focus on regions as level of analysis makes them poorly placed to explain network phenomena such as social capital and learning (Moulaert & Sekia, 2003; Oerlemans et al., 2007). Second, the view of regions as nodes in global knowledge creation networks questions some of the key assumptions regarding the spatial dimension of knowledge that underlies the TIM literature. By departing from a more or less explicit understanding of territory (Oerlemans et al., 2007), the TIM literature, on the one hand enables comparison between regions, but on the other hand insists on an increasingly problematic position that local knowledge production prevails over global knowledge production (Malecki, 2010).

In sum, the TIM literature's understanding of social capital and its effect on learning can only be superficial as it fails to grasp the obvious fact that both social capital and learning are relational concepts. The role of social capital in regional learning networks should therefore be approached from a relational rather than a territorial perspective. This requires a different turn in economic geography, away from the traditional focus on regions (and firms), and toward a focus on (networks of) individuals.

SUMMARY AND CONCLUSION

This chapter set out to highlight the role of social capital with respect to learning in regional networks. Social capital may be defined as trust, norms, values, and customs, socio-cultural diversity, etc., and plays an important role in regional learning because learning essentially is a social process among individuals. It is a matter of social interaction rather than economic transaction, as it is often portrayed in much of the mainstream economic and economic geography literatures. Even though this literature has gradually recognised the role of social capital, it has thus far not successfully conceptualised this role. The TIM literature which addresses the relation between space and learning was also found inadequate in its conceptualisation of both learning and social capital. This chapter discussed the role of the social capital in regional networks from several perspectives with the ambition to draw some conclusions as to how this role may be conceptualised. The follow key findings have emerged:

1. Social capital is foremost a characteristic of networks. Social capital shapes the social interactions taking place in networks, such as learning, in several ways. Since networks may be seen as the vehicles for social interaction between individuals, so is social capital a characteristic of these networks as they are enacted in them. The key issue for learning in regional networks is that regional or societal social capital can facilitate economic interaction such as learning. This is illustrated by the Western Java case where such a crossover of societal social capital to economic relations is largely absent, which seems to hamper regional learning.
2. Some forms of social capital are more conducive to learning than others. In general, networks possessing social capital that encourage openness, tolerance, socio-cultural diversity, and creativity fare better at learning than other networks. Moreover, networks that have both high levels of bonding and bridging social capital are more conducive to learning than other types of networks.
3. Since the individuals populating networks are spatially sticky, so is social capital. The obvious question then arising—why are individuals spatially sticky?—may be answered by the fact that certain forms of social capital that are beneficial to learning attract people to some places rather than other ones. However, empirical research supporting this argument is still thin.
4. The spatial scale of learning in networks ranges from local to global. Even networks that are strongly embedded in a particular region are more often than not strongly linked to global knowledge networks as well. That is because 'knowledge assets' are much less territorially defined as the TIM literature suggests. This further underlines that social capital is best seen as characteristics of networks rather than of regions as it is difficult to see how specific regional social capital can affect global knowledge networks. It is more likely that the social capital of a network comes from multiple sources brought into the network by the various individuals populating them.

5. Looking at social capital at the level of regions presents other problems. It basically turns the question why and how specific regional social capital affects learning in networks into a black box. Moreover, at this point the TIM literature may be biased toward industrial regions where social and economic networks overlap and region-specific social capital may thus easily permeate to economic life. This may not be the case in regions in developing countries, which further troubles the conceptual link between regional social capital and learning in networks.

The conclusion of this chapter is that the effect of social capital on learning is best studied on the level of individuals in regional networks. That is, networks of individuals that are embedded in a specific location but may have a substantial number of non-regional or even global members. Because social capital is enacted in the social interactions between individuals and because learning is a social process between individuals (rather than an economic transaction), the level of individuals in regional networks seems the most appropriate level to study the role of social capital in regional learning processes. Looking at regional networks also enables a more sophisticated explanation, compared to the black box of the TIM literature, of how regional social capital affects learning in networks and why people and social capital may be spatially sticky. The present chapter has presented some preliminary answers to this. Turning back to the questions at the start of this chapter—why do some regions, or rather regional networks, perform better in the global economy—this chapter suggests the following. It is necessary to look, first, at how effective network and societal (regional) social capital facilitates learning among individuals in regionally embedded networks, and second, at the extent to which regional social capital succeeds in shaping an environment that is conducive for creativity and learning and that draws and binds people to the region. This requires focusing on (networks of) individuals to a far greater degree than mainstream economic geography has thus far exhibited. In the light of the failure of the TIM literature to convincingly conceptualize the relation between space and learning and the role of social capital in it, this 'individual turn' in economic geography seems a more productive way forward.

NOTES

1. See *www.oica.net/category/economic-contributions/auto-jobs*. Accessed December 1, 2009
2. Sources: Statistics Indonesia; Statistics Jakarta; and Badan Pusat Statistik Provinsi Jawa Barat. See also *www.dds.bps.go.id*. Accessed December 1, 2009
3. Source: Gaikindo (The Association of Indonesian Automotive Industries). ASEAN refers to Thailand, Malaysia, Indonesia, Philippines, Vietnam, and Singapore.
4. Source: Gaikindo.

REFERENCES

Amin, A., & Cohendet, P. (2004) *Architectures of knowledge: Firms, capabilities and communities*, Oxford: Oxford University Press.

Best, M. (1990) *The new competition: Institutions of industrial restructuring*, Cambridge: Polity Press.

Beugelsdijk, S., & Van Schaik, T. (2005) Differences in social capital between 54 Western European regions. *Regional Studies*, 39 (8), 1053–1064.

Breschi, S., & Malerba, F. (2005) *Clusters, networks, and innovation*. Oxford: Oxford University Press.

Burt, R. (2005) *Brokerage and closure: An introduction to social capital*, Oxford: Oxford University Press.

Busser, R., &d Sadoi, Y. (eds.), (2004) *Production networks in Asia and Europe: Skill formation and technology transfer in the automobile industry*. London: Routledge.

Chen, M. (1996), *Managing international technology transfer*. London: International Thomson Business Press.

Doner, R. (1991) *Driving a bargain: Automobile industrialization and Japanese firms in Southeast Asia*. Berkeley: University of California Press.

Dupuy, C., & Torre, A. (2006) Local clusters, trust, confidence, proximity. In Christos, P., Sugden, R., & Wilson, J. (eds.), *Clusters and globalization: The development of urban and regional economies*. Cheltenham: Edward Elgar.

Durkheim, E. (1893/1997) *The division of labor in society*. New York: Free Press.

Field. J. (2003), *Social capital*. London: Routledge.

Florida, R. (2002) *The rise of the creative class, and how it's transforming work, leisure, community and everyday life*, New York: Basic Books.

Gerlach, M., & Lincoln, J. (1992) The organization of business networks in the United States and Japan. In Nohria, N., & Eccles, R. (eds.), *Networks and organizations: Structure, form and action*. Boston: Harvard Business School Press.

Grabher, G. (2004) Temporary architectures of learning: Knowledge governance in project ecologies. *Organization Studies*, 25 (9), 1475–1489.

Gwyne, R. (1990) *New horizons? Third world industrialization in an international framework*. London: Longman.

Hassink, R. (2001) The learning region: A fuzzy concept or a sound theoretical basis for modern regional innovation policies? *Zeitschrift für Wirtschaftsgeographie*, 45 (3/4), 219230.

Hatch, W., & Yamamura, K. (1996) *Asia in Japan's embrace: Building a regional production alliance*. Cambridge: Cambridge University Press.

Hauser, Ch., Gottfried, T., & Walde, J. (2007) The learning region: The impact of social capital and weak ties on innovation. *Regional Studies,*. 41 (1), 75–88.

Honda. (2004) *World motorcycle: Facts and figures*. Tokyo: Honda Motor Co. Ltd.

Hudson, R. (1999) The learning economy, the learning firm and the learning region: A sympathetic critique of the limits of learning. *European Urban and Regional Studies*, 6 (1), 59–72.

Huntington, S., & Harrison, L. (eds.). (2000) *Culture matters: How values shape human progress*. New York: Basic Books.

Inglehart, R., & Baker, W. (2000) Modernization, cultural change, and the persistence of traditional values. *American Sociological Review*, 65 (1), 19–51.

Irawati, D. (2009) *The Indonesian automotive cluster and its relationship in the global production network of the Japanese multinational enterprises: A case study of Toyota and Honda*. PhD Thesis, Newcastle: Newcastle University.

Irawati, D., & Charles, D. (2010) The involvement of Japanese MNEs in the Indonesian automotive cluster. *International Journal of Automotive Technology and Management*, 10 (2/3), 180–196.
Lagendijk, A., & Charles, D. (1999) *Clustering as a new growth strategy for regional economies?* Paris: OECD.
Larson, A. (1992) Networks dyads in entrepreneurial settings: A study of governance in exchange relations. *Administrative Science Quarterly*, 37 (1), 76–104.
Lazerson, M., & Lorenzoni, G. (1999) The firms that feed industrial districts: A return to the Italian source. *Industrial and Corporate Change*, 8 (2), 235–266.
Lorenzen, M. (2007) Localized learning and social capital. In Rutten, R., & Boekema, F. (eds.), *The learning region: Foundations, state of the art, future.*, Cheltenham: Edward Elgar, pp. 206–230.
Lorenzen, A. (2008) Knowledge networks in local and global space. *Entrepreneurship & Regional Development*, 20(6), 533–545.
Malecki, E. (2010) Global knowledge and creativity: New challenges for firms and regions. *Regional Studies*, 44 (8), 1033–1052.
Markusen, A. (1999) Fuzzy concepts, scanty evidence, policy distance: The case for rigour and policy relevance in critical regional studies. *Regional Studies*, 33 (9), 869–884.
Marshal, A. (1890/2006) *The principles of economics.* New York: Cosimo.
McFadyen, A., Semadeni, M., & Cannella, A. (2009) Value of strong ties to disconnected others: Examining knowledge creation in biomedicine. *Organization Science*, 20 (3), 552–564.
Morgan, K. (1997) The learning region: Institutions, innovation and regional renewal. *Regional Studies*, 31 (5), 491–503.
Morgan, K. (2004) The exaggerated death of geography: Learning, proximity and territorial innovation systems: *Journal of Economic Geography*, 4 (1), 3–21.
Morgan, K., & Nauwelaers, C. (eds.). (2003) *Regional innovation strategies: The challenges for less-favoured regions.* London: Routledge.
MOSR, (2002) *The industrial and research policy of Indonesia.* Jakarta: Ministry of Science and Research.
MOTIRI, (2005) *The automotive industry and its progress,* Jakarta: Ministry of Trade and Industry.
Moulaert, F., & Sekia, F. (2003) Territorial innovation models: A critical survey. *Regional Studies*, 37 (3), 289–302.
Nooteboom, B. (2006) Innovation, learning and cluster dynamics. In Asheim, B., Cooke, P., & Martin, R. (eds.), *Clusters and regional development: Critical reflections and explorations.* Abingdon: Routledge, pp. 137–163.
Oerlemans, L., Meeus, M., & Kenis, P. (2007) Regional innovation networks. Rutten, R., & Boekema, F. (eds.), *The learning region: Foundations, state of the art, future.* Cheltenham: Edward Elgar, pp. 160–183.
Ozawa, T. (2005) *Institutions, industrial upgrading and economic performance in Japan: The flying geese paradigm of catch-up growth.* Cheltenham: Edward Elgar.
Porter, M. (1990) *The competitive advantage of nations.* New York: Macmillan.
Pries, L., & Schweer, O. (2004) The product development process as a measuring tool for company internationalization. *International Journal of Automotive Technology and Management*, 4 (1), 1–21.
Rugraff, E. (2010) Foreign direct investment (FDI) and supplier-oriented upgrading in the Czech motor vehicle industry., *Regional Studies*, 44 (5), 627–638.
Rutten, R., & Boekema, F. (2007) The learning region: A conceptual anatomy. In Rutten, R., &d Boekema, F. (eds.), *The learning region: Foundations, state of the art, future.* Cheltenham: Edward Elgar, pp. 127–142.

Rutten, R., & Gelissen, J. (2010) Social values and the economic development of regions. *European Planning Studies*, 18 (6), 921–940.

Rutten, R., & Oerlemans, L. (2009) Temporary inter-organizational collaboration as a driver of regional innovation: An evaluation. *International Journal of Innovation and Regional Development*, 1 (3), 211--234.

Rutten, R., Westlund, H., & Boekema, F. (2010) The spatial dimension of social capital. *European Planning Studies*, 18 (6), 863–870.

Scott, A., & Storper, M. (2003) Regions, globalization, development. *Regional Studies*, 37 (6 & 7), 579–593.

Storper, M. (1993) Regional worlds of production: Learning and innovation in the technology districts of France, Italy and the USA. *Regional Studies*, 27 (5), 433–455.

Storper, M., & Venables. A. (2004) Buzz: Face-to-face contact and the urban economy. *Journal of Economic Geography*, 4 (4351–370.

Teece, D. (2000) Strategies for managing knowledge assets: The role of firm structure and industrial context. *Long Range Planning*, 33 (1), 35–54.

Torre, A. (2008) On the role played by temporary geographical proximity in knowledge transmission. *Regional Studies*, 42 (6), 869–889.

Toyota. (2007) *Toyota annual report*. Tokyo: Toyota Motor Corporation.

Uzzi, B. (1997) Social structure and competition in interfirm networks: The paradox of embeddedness. *Administrative Science Quarterly*, 42 (1), 35–67.

Westlund, H., & Adam, F. (2010) Social capital and economic performance: A meta-analysis of 65 studies. *European Planning Studies*, 18 (6), 893–920.

Part III
Case Studies

8 Independent Inventors and their Position in the National/Regional Innovation System
A Case Study of Slovenia

Angela Ivančič, Darka Podmenik, Alja Adam, and Ana Hafner

INTRODUCTION

Various historical texts and scientific sociological literature suggest that throughout history individual inventors have played an important role in technological development (see Wagner, Weick, & Eakin, 2005). In contrast, they are today seriously undervalued and marginalised as a research subject. The relatively rare studies chiefly address the contribution of individual inventors to economic and technological development relative to corporate inventors in terms of their contribution to the body of patents (Nicholas, 2011; Wagner, Weick, & Eakin, 2005; Anderson, 2004), or the technological complexity of the patents and their future influence as measured by citations (Dahlin et al., 2004; Wagner, Weick, & Eakin, 2005; Fleming, 2006; Wuchty et al., 2007; Singh & Fleming, 2010; Lettl et al., 2009). Such research findings mostly confirm the diminishing role of independent inventors compared to corporate inventors.

Fleming (2006) reports that independent inventors differ from their corporate counterparts in that they are more successful in generating new combinations of knowledge when they solve problems related to well-defined technologies. In contrast, Dahlin et al. (2004) showed that independent inventors are such a diverse population that patents with the greatest impact on the population belong to them, although at the same time they are more likely than corporate inventors to patent inventions that have little impact.

Several contemporary studies point to a shift from the "lone inventor" model to the teamwork model (Wuchty et al., 2007). It is argued that teamwork enables a better selection of ideas which results in fewer failures. Singh and Fleming (2010) propose that collaboration can have opposite effects at two extremes: it reduces the number of very poor outcomes while increasing the probability of extremely successful outcomes because of the greater recombinant opportunity in creative search.

Lettl et al. (2009) assert that independent inventors have less absorptive capacity than organisations and the individuals operating within them.

Organisations can act collectively on new technological knowledge and, by engaging in practical co-operation, convert that knowledge into technological knowledge relevant for future action.

Fuglsang (2008) argues that innovation simultaneously requires diversity and collectivity, as well as a balance between the two, which is affected by market, organisational, and social forces. This raises the question of innovation structures of the external environment and their characteristics that can either support or inhibit innovation activities at the organisational and individual levels. The concepts of institutions, learning, and collaboration are important in the discourse on innovation. The literature on national innovation systems[1] argues that the way in which interrelations are structured among institutions and actors in the innovation system in a particular country significantly affects innovation performance (OECD, 1997; Westlund, 2011; Adam et al., 2010).[2] The literature also underlines the importance of regional innovation systems characterised by co-operation in innovation activity between firms and knowledge creating and diffusing organisations, such as universities, training organisations, R&D institutes, technology transfer agencies, and so on, and an innovation supportive culture that enables both firms and systems to evolve over time. Cooke (2001) highlights the significance of promoting interactions between different innovative actors who expect advantages from them. These interactions may embody localised interactive learning but also include the wider business community and governance structure.[3]

At the turn of the century, the social capital approach to the study of innovation gained particular importance. In their overview of knowledge-based theories of innovation, Landry et al. (2002) show that the theories, questions, and approaches have changed drastically over the past 40 years—from technical to social-relational. Trust and established networks are emphasised as positively influencing inventions and creativity (Knack & Keefer, 1997; Paxton, 1999). However, research dealing with independent inventors only rarely tackles this issue (see Lettl et al., 2009; Wuchty et al., 2007).

This study aims to contribute to the existing body of research in this field by investigating two dimensions of the innovation activities of independent inventors—entrepreneurs and individual inventors (sometimes called "hobbyists"), namely their ability to:

- generate social capital as a factor facilitating their innovation activities; and
- identify support institutions and mechanisms in the innovation environment at the national and local levels and establish collaborative relations with those institutions in order to make better use of their services, and thereby maximise resources facilitating innovation activities.

We build on the assumption that the ability to successfully apply social capital accumulated in building collaborative networks can enable independent inventors to be more successful in coping with deficiencies resulting from their marginalised position in the innovation system and from the absence of the organisational resources corporate inventors enjoy. In collaborative networks independent inventors can share knowledge and information, and recognise and utilise opportunities provided by the formal and non-formal innovation environment at national and local/regional levels.

Taking into consideration the literature arguing that innovation and technology are gendered areas (Danilda & Granat Thorslund, 2011), we also wish to explore these issues from the gender perspective, which represents another neglected issue in the existing research on independent inventors.

The study is based on a case study of Slovenia. Slovenia is an interesting case since recent statistical data regarding the R&D field indicate a considerable improvement of its position among EU countries (EC, 2011). At the same time, national evidence shows that about 40% of all national patents granted in the application year 2007 were granted to independent inventors (Adam et al., 2010). Independent inventors are thus an important agent of the creation and diffusion of knowledge.

The chapter is structured as follows: the introduction is followed by the theoretical framework of the study which presents the central theoretical foundations and main theories used for conceptualising the study. We primarily rely on theories stressing the role of social capital and social networks in the process of innovation. We then describe the methodological approach to the study and, finally, the results of the analysis are presented and discussed.

THEORETICAL CONCEPTUALISATION

Two main approaches to social capital have been identified as conceptually adequate for explaining endowment with new knowledge and inventive solutions: the structural approach and the individual approach (see Masciarelli, 2011). Based on the premise of a "collective good," the former accentuates the importance of participation in institutions and organisations at different levels of the social structure: in local social interaction (Coleman, 1988); regional governmental (Putnam, 1993; Hauser et al., 2007) and provincial units (Guiso et al., 2004) as well as national entities (Knack & Keefer, 1997). The individual approach is based on the premise that the "private good" motivates the accumulation of social capital. From this point of view, social capital is seen as an individual investment in social networking which, according to P. Bourdieu, one of the founding fathers who developed the concept of social capital in the 1970s and early 1980s (Siisiäinen, 2000), allows access to different sorts of resources and can contribute to different sorts of profits.

Several authors produce evidence that social capital in terms of trust and established networks positively influences inventions and creativity (see Knack & Keefer, 1997; Paxton, 1999). Both forms of social capital—structural social capital and individual social capital—have proved to be important. Social capital can be analysed as an individual's asset, but also an asset of a community, region, or firm (Kaasa, 2007, 6).

According to Putnam (1993), social capital is a feature of civil society and its structures. However, following Westlund (2006; 2011), besides civil society social capital also resides in institutions (industry, public sector bodies, and other organisations). Social capital therefore exists in innovation systems as well. It is presumed that a difference between independent and corporative inventors also exists in the type of social capital they possess: independent inventors are expected to primarily use social capital accumulated in civil society networks whereas corporate inventors have organisational social capital available which provides them with opportunities to collaborate in various formal and informal networks composed of different innovation actors that operate within the innovation system. To make up for such deficiencies, independent inventors need abilities and resources to also gain access to the social capital available in formal institutions.

Extensive research demonstrates that social capital provides several advantages with regard to acquiring new knowledge and establishing paths for knowledge spillovers. It creates learning opportunities (Powell et al., 1996); facilitates knowledge sharing (Bourdieu, 1986; Coleman, 1988); increases interpersonal and social trust (Putnam, 1993) and facilitates joint problem solving (Lettl et al., 2009). It also enables the transformation of information into knowledge[5] and supports the convergence of many kinds of knowledge held by different categories of actors[6] (Landry et al., 2002).

In addition, the literature on innovation systems and regional science literature conceptualises innovation as an evolutionary and social process in which social capital plays an important role (Edquist, 2004; Cooke et al., 2000). It is argued that innovation is stimulated and influenced by many actors and factors, both internal and external to the firm. The social aspect of innovation refers to the collective learning process among several departments of a company (for example, R&D, production, marketing, commercialisation, etc.) as well as to external collaboration with other firms, knowledge providers, finances, training, etc. (Cooke et al., 2000).

Case studies of regional innovation systems conducted so far have focused on the importance of trust and co-operation in networks and the way in which positive social relations produce information sharing and knowledge spillovers (cf. Cooke, 2005). According to Wolfe (2002), the existence of social capital and trust as an element of social capital helps overcome market failures or reduce market costs for firms in densely related networks by supporting stable and reciprocal exchange relationships. However, some authors point out that trust and co-operation among co-located firms is limited (Freel & Harrison, 2006). De Filippis (2001) notes in his critique that networks of all kinds are constructed around relative power relations, while Lin et al. (2001) suggest that some positions in a network carry more valued resources and provide options for the exercise of greater power. Networks can and frequently do take the form of hierarchies where less powerful members enjoy only marginal benefits.

Although the literature on innovation systems as a rule marginalises independent inventors as innovation agents, it may be assumed that irrespective of their formal invisibility they are embedded in the system as either entrepreneurs or members of civil society who interact with other actors in the innovation system as members of various communities or user-inventors. Hence it seems justified to extend the above arguments to this group of innovation producers.

RESEARCH QUESTIONS AND METHODOLOGY

The study aims to explore the efficiency of independent inventors in accumulating different forms of social capital and building networks as important resources supporting their innovative activities, and the extent to which they are able to identify and utilise provisions of the external environment supporting innovation at the national and regional/local levels. Can one assert that effective utilisation of these opportunities and their effective transformation into resources helps them overcome their marginal position in the system?

The following questions are addressed:

- Which are the main characteristics of social capital generation and networking? Which types of social capital are prevalent in the case of Slovenian independent inventors and how do they evaluate their contributions to (national) inventive society?
- How do independent inventors perceive their position within the national/regional innovation system and their relations to structures acting as agents supporting innovation processes at the national and regional/local levels?
- To what extent do independent inventors utilise the opportunities provided by external environments supporting innovation at different levels?
- What is the role of accumulated social capital and established networks of independent inventors in identifying and utilising the provisions offered by supporting agents at different levels in maximising their resources?
- Are there gender-related differences in independent inventors' accumulation of social capital and their access to what is provided by the supportive national/regional/local environment?

Methodological Approach

The qualitative methodology is adopted with a focus on a case study of Slovenia. The case study incorporates secondary data analysis and semi-structured, in-depth interviews with independent inventors in Slovenia. A secondary analysis of the literature is used to review formal documents (legislation, policy documents, official reports) and other contributions presenting research findings and analyses of statistical data. Interviews were conducted to collect individual-level information.[7]

The prime criterion for selecting the interviewees was that a person had been granted at least one patent or industrial design. Respondents were selected from the list of independent inventors maintained by the Slovenian association of inventors called ASI (Active Slovenian Inventors) which supports and connects independent inventors. While selecting the respondents, we stove towards purposeful representation (Densin & Lincoln, 1994) rather than empirical representation. Both types of independent inventors—entrepreneurs and individual inventors—were included.[8]

Respondents were selected within two age cohorts—those 50 years of age and over and those below 40 years of age. These age cohorts were chosen because the innovative environment in Slovenia has changed considerably in the past three decades when the country started to experience a shift toward a knowledge-based economy and information technology which has significantly affected all domains of life. This suggests different approaches in coping with the demands characterising innovation processes especially with regard to ways of accumulating social capital and networking.

Since we wanted to investigate gender differences, females are overrepresented in the sample.[9] Altogether, 22 independent inventors were selected. The majority (17) are males and five are females. In age they range from 23 years to 92 years with a mean age of 55.6 years.[10]

Individual face-to-face interviews were carried out. We opted for topic-focused, semi-structured interviews. An interview protocol was prepared that consisted of open-ended questions. Most interviews were carried out in July and August 2010. In order to ensure a more balanced sample with regard to gender and age, eight additional interviews were conducted in August 2011.

All of the interviews were taped and transcribed, transcriptions are available at the Social Data Archive (http://adp.fdv.uni-lj.si/; accessed January 30, 2012).

MAIN CHARACTERISTICS OF SOCIAL CAPITAL CREATION AMONG SLOVENIAN INDEPENDENT INVENTORS

As argued above, independent inventors are put on the margins of the 'system-wide processes' which are keeping innovation alive. Accordingly, they are deprived of the wide range of resources available to their employed colleagues, such as: opportunities for bridging diverse technological fields; being part of the corporate intelligence and/or supportive organisational frameworks with their corresponding information resources. On the other hand, organisational competencies do not restrict and burden their inventiveness and creativity. It is evident from our empirical evidence that independent inventors are an inexhaustible source of inventive ideas but, because of limited resources and an absence of adequate support from external institutions, it is practically impossible to further develop and exploit many of them.

According to the literature, inventive activities can be divided into two main phases: innovation initiation and innovation implementation (Glynn, 1996, in Plankerhorn, 2008, 99). In the first phase, new ideas are generated and presented in a form understandable to others. In the second phase, inventive ideas are adopted, developed, and finally exploited. The first phase is more individualistic and independent of the social environment, while the second requires various sorts of environmental support and co-operation among different actors on different institutional levels. Regarding the independent inventors also defined as those who are 'lonely,', marginalised or 'hobbyists,' it could be posited that during the first phase they feel like fish in water, while in the second phase they feel just the opposite.

The main question considered below is whether independent inventors are capable of using social capital and human capital as 'substitutes'« for the system-wide deficiencies to bring their inventive ideas to life.

INDEPENDENT INVENTORS' STRUCTURAL AND INDIVIDUAL SOCIAL CAPITAL

The results of our analysis confirm the marginalisation of the interviewed inventors. They believe the national innovation support system should provide more resources and backing for them. They express overwhelmingly negative and critical opinions of the national and regional institutional environment. More experienced respondents seem to be more critical of the way in which the status and rights of independent inventors are managed at the national and systemic level than are their younger counterparts.[11] Regarding their criticism, it may be concluded that the national-level networks are not functioning adequately. Institutional support at different levels is suggested to be insufficient precisely in the innovation process phase where it is most needed, namely, during the transfer from the first to the second phase of inventive activities: from initiation of the innovation to its implementation.

The interviews suggest that the lack of structural social capital (Masciarelli, 2011) in the first phase of invention creation could be compensated with 'individual' resources, either familiar or associational (civil societal). Especially younger respondents obtain different kinds of support from their families, unlike the older ones who mainly draw support from a national association. But the statements of both point out the weakness of the structural level of building social capital, especially on the contact points between so-called supportive institutions and the expectations of independent inventors with regard to institutionalising their inventive ideas.

Some significant differences between the younger and older interviewees are evident in our research. The networks of the younger ones are more widespread, more professional and heterogeneous. They often use the Internet to gain access to new knowledge and information while the older inventors (above 65 years of age) are usually not sufficiently e-literate to enjoy this advantage. The younger ones tend to aim at entrepreneurship at the very beginning of the innovation or in an early phase and are more open to exchanging knowledge and assessment concerning their ideas.

Considering individual networking as a form of compensation which should contribute to the creativity and successfulness of independent inventors, the family as a supportive environment in the case of our respondents proved to be significant. With regard to the younger inventors' inventiveness it could be even said that the human and social capital of their parents plays an important role.[12]

Most respondents were already inclined to innovation in their childhood. The majority also said their parents did not support them directly but that they instead tolerated their early attempts at innovation. As R 14 recalls, *"Both my parents have positive attitudes to innovations . . . They did not inhibit me or my brother in our pursuits and left us to our own devices . . for example, we were allowed to remodel the furniture with wallpaper."* (R 14).

UPGRADING INDEPENDENT INVENTORS' KNOWLEDGE AND SKILLS

The majority of the respondents achieved a higher education level than their parents and therefore a direct transfer of intergenerational human capital in the family framework should not be expected.[13] However, some respondents suggest that parents can be an important source of tacit knowledge built up via mutual understanding and the sharing of everyday experiences. One respondent noted that as a child she was inspired by her father, who *"was always conjuring up and creating something around us."* (R 11). As a supporting environment, besides the family, high school and workplaces are mentioned the most frequently in the interviews. The respondents do not invest much in the gaining of additional knowledge and skills via externally organised courses; those who participate in further education are mostly practically oriented professionals. The data confirm the above-mentioned proposition concerning independent inventors' exclusion from formalised supporting networks.

Such exclusion from the innovation system and knowledge production makes professional and cross-disciplinary ties a weak point of independent inventors and thus not a factor compensating for the missing organisational support. Some authors suggest that specialisation in a given technological field may contribute to their professional networking, as well as their embeddedness in a community of peers (Brown & Duguid, 2001). Such communities may provide independent inventors with a social context for learning and creativity (Perry-Smith & Shalley, 2003), for the filtering and selection of promising ideas (Von Hippel, 2007) and for disseminating inventions (Fleming, 2007). Among professional organisations, the national association of Slovenian independent inventors (ASI) is mentioned most often as a source of different types of support. With regard to exchanging ideas and getting new ones, innovation fairs (more frequently called exhibitions of inventions)[14] are the most attractive professional events. The interviewees also mentioned the educational role of fairs, most vividly described by one of them: *"I learnt how to present and also how to sell a product . . . I also learnt how to deal with those who are only interested in your brochure and real customers, and how to deal with someone who only wants a chat."* (R 6)

Establishing and maintaining professional networks seems to depend on the age of the interviewed independent innovators. We found that the younger inventors are more likely to plan and maintain professional networks. According to one of them, *"Yes, I have devoted much time to this and have developed a good network. . . . I tried to attend most innovation events in Slovenia and I met a large number of people. I also made some new acquaintances at the Chamber of Commerce and Industry . . . and slowly I have come full circle."* (R 14)

However, the interviewees believe that friendships networks are relatively insignificant for their inventive activity and as a reason for such 'unusefulness' they point out their friends' disinterest in professional inventions matters. In addition, the unimportance of "non-professional" contacts is reflected in the interviews where the respondents devote so much attention to describing their practices, experience, and achievements. As one inventor states, *"I am of course communicative when issues that are common to me are being discussed, that is at certain conferences where people with similar interests meet. In a more general environment, I am more introverted or, as my wife puts it, boring."* (R 5)

TRUST IN OTHERS AND INSTITUTIONS

In the conceptual framework of social capital, uncertainty avoidance is negatively linked to the degree of social trust, meaning that stronger mechanisms for uncertainty avoidance result in less social trust and less inventive ideas and actions taken. The level of trust in society is also important. The greater the level of trust, the more likely people are to enter co-operative relationships with others. Conversely, when the level of trust in society is low, people are generally only willing to co-operate with partners with whom they have close contact (Dakhli & De Clercq, 2004).

Innovations are associated with different types of change[15] and some kind of uncertainty. Risk-averse attitudes mean there is less incentive to come up with new ideas and, even if such ideas are proposed, they are more likely to be rejected. On the other hand, there is a stronger tendency in such environments to protect intellectual property through patenting (Kaasa & Vadi, 2008, 9). Our empirical evidence shows there is a negative relationship between uncertainty avoidance and innovation initiation.

The interviewees seem to have built their safety niches with primary networks; they became sole entrepreneurs or opened small family enterprises; they look for information and exchange ideas in civil society associations. But they are generally very distrustful, especially the older, more experienced ones who have battled to protect their patents for many years.

The low level of trust is particularly reflected in the respondents' unwillingness to work in teams. Nevertheless, they usually have to co-operate with different people such as sole traders and engineers in a variety of fields in the process of manufacturing prototypes and in later stages. They also have to co-operate with administrators and managers when acquiring and protecting their patents. The majority of our respondents see such collaboration as "unavoidable" and are aware they must be very careful when selecting their collaborators. Some of them have had bad experiences, such as their ideas being stolen by a sole trader making the prototype or by a mediator who was supposed to lobby for the invention.

There are differences in the younger and older respondents' experiences and one of them is that younger inventors have a higher level of trust and willingness for teamwork than the older ones.

THE SUCCESSFULNESS OF INDEPENDENT INVENTORS—ECONOMIC SUCCESS IS NOT THE MAIN INDICATOR

In terms of financial performance, most innovators "daydream" of the profitability of their innovations and personal wealth, although money is not the main motivator of their activities. A survey of American independent inventors (Whalley, 1992) revealed that the majority see innovation as a challenge and/or source of pleasure and only half of them are motivated by money. The rest feel that innovation is a search for solutions to problems in their primary work or social environment. If financial success is not the fundamental reason for independent inventors' efforts, then their success should not only be measured in terms of market success or material profits. More than half the Slovenian inventors who participated in this research have made no money from their innovations, even though they all hold patents or industrial designs and awards for their innovative achievements. Less than half of them have earned an average of between EUR 500 and EUR 3,000 a year. Slightly less than one-third of the respondents feel they are successful despite their inventions earning no income, and as many as two-thirds feel they are successful or even very successful regardless of their earnings.

Based on the results of the Slovenian qualitative study, discussions with members of the management of innovative organisations, and written materials, we may conclude that independent inventors develop some kind of autonomous, independent criteria for assessing their inventions and the success of their colleagues' inventions. The life stories told by these inventors show that, in most cases, they have systematically pursued creative innovative activities inbred in them since early childhood. We could call it a talent (not unlike the talent of artists) which cannot be denied without major personal sacrifice. One of our respondents describes the origin of his latest invention, *"Maybe it is hard to believe, but the idea came to me in my sleep. I woke up and saw things very clearly. I said to my girlfriend: 'Just a minute, I need to draw this,' I took a pencil and drew an illustration . . . I have already sent the patent application and we are currently discussing the sale of the patent to a manufacturer."* (R 22)

Nearly all of the interviewees stress the importance of each individual's creative freedom and do not believe in teamwork. As described by R 16 who originally worked in co-operation and with the support of her husband but has since become independent, *"Everyone must decide for themselves whether they prefer security or freedom, and this takes some effort. ."*

THE NATIONAL/REGIONAL SUPPORTIVE ENVIRONMENT FOR INNOVATION AND INDEPENDENT INVENTORS

One of the questions guiding our exploration addresses the access of independent inventors in Slovenia to institutional support provided by the external environment at the national/regional levels. We explore the extent to which measures and activities provided by supportive agents within the national/regional innovation system extend to independent inventors as well as the interviewees' perception of the institutional support and their approach to exploiting it.

In Slovenia a large and growing number of institutions with very different kinds of specialised expertise are involved in the production and diffusion of knowledge. However, the number and variety alone do not guarantee the quality of these institutions' work. According to Adam et al. (2010, 171–174), the supportive environment institutions are too bureaucratic to be efficient in their provision of individual supportive measures and programmes. While discussing independent inventors it has to be strongly emphasised that most organisations declared as being an element of the supportive environment for innovation are mainly part of the supportive environment for entrepreneurship and that individual inventors are excluded from enjoying the benefits of their support.[16] It can be concluded that independent inventors interested in establishing their own enterprise have access to counseling, information, and training activities as well as certain financial mechanisms facilitating start-ups. However, those not interested in becoming entrepreneurs are mostly excluded from such support.

FINDINGS FROM THE INTERVIEWS: INSUFFICIENT SUPPORT

Most inventors in our sample were individual inventors at the time the interview was conducted, although three of them had experience with entrepreneurship or were craftsmen earlier in their lives. Among those who were entrepreneurs, eight owned enterprises that were connected with their inventions, while in two cases the enterprises had nothing to do with the respondents' inventions (one was a freelance photographer, for instance, but his invention was not from this field).

In our analysis we focused on the few supportive organisations we assumed to be the key institutions of the national supportive environment for innovation and which extend their services to all inventors: public research organisations (PROs), technology parks (TPs), and business incubators (BIs), the Slovenian Intellectual Property Office (SIPO), and inventors' associations. In general, our interviewees expressed quite negative opinions about the activities of these institutions. This is especially the case of the older cohort. In contrast, on average the younger inventors typically had less experience with supportive organisations, which is understandable since they are at the beginning of their innovative paths.

The highest number of negative opinions involved co-operation with TPs and BIs, which is very curious since their primary mission in Slovenia is to stimulate regional entrepreneurship, especially technological and innovative start-ups, among inventors whose ideas are suitable for commercialisation.[17] Respondent R 2 asserts "*I do not underestimate those institutions ... however I fear that they mostly spend the money they get from the state on themselves and not on offering assistance to inventors.*"

Their estimation of co-operation with PROs was also quite negative or undefined, with some important exceptions who reported successful co-operation. However, a good number of interviewees would seek an expert opinion from scientists and researchers when developing a prototype. Co-operation with PROs is also desired at the stage of commercialising an invention. In the last two years, four technology transfer offices offering professional help in licensing inventions and selling patents have been established in Slovenia, yet their services are mainly available to the inventors employed by PROs. Moreover, independent inventors in Slovenia cannot obtain professional and free-of-charge assistance with licensing their patents. It is thus understandable that when inventors were asked about the most desirable new supportive organisation that could be established in Slovenia, a technology transfer office for independent inventors was mentioned.

Co-operation with the SIPO was not often mentioned, even though almost all inventors in the sample had at least one patent, industrial design, or trademark granted, and at least some contact with the SIPO was necessary for those who had not filed their patent applications through patent attorneys. Some inventors also expressed their dissatisfaction with the SIPO, although their criticism was not directed so much at its staff but toward the patent system itself. A typical statement here was made by respondent R 5, *"Patents only protect the rich."*

According to our research findings, associations of inventors seem to be one of the most supportive institutions for independent inventors. Their main purpose is to promote inventions and inventors and support the sharing of experience among members. Most countries around the world have at least one association of inventors, set up as a non-governmental organisation, although in some countries like Germany, Croatia, or Finland there are several. Many national associations are members of the International Federation of Inventors' Association (IFIA), which is active in promoting inventions as well as in striving to achieve better conditions regarding independent inventors all around the world (cheaper patent protection, fighting against inventor scams, encouraging female inventors, etc.). For example, small entry fees for individual inventors and small companies were (unsuccessfully) proposed in all countries. In the past 10 years in Slovenia, most of the many former associations of inventors established at the level of regions and municipalities have disappeared. The inventors' response to this was to form a new, state-wide association named ASI (Active Slovenian Inventors) in 2005 (Podmenik & Hafner, 2011).

However, a lot of inventors see the association as "the only supportive environment." Some even believe that membership in the association should directly lead to their commercial success and are therefore disappointed when such expectations do not materialise (ibid., 148, 149).

Our findings are in line with the findings of Adam et al. (2010) showing that, in addition to the lack of institutional and organisational support on the transition from the first to the second innovative phase, the first phase itself is also not successfully supported. Applying for a patent and making the first prototype is a burden on inventors' shoulders alone and they cannot expect much support from institutions. As one interviewee observed, independent inventors must first gain recognition from others, usually in the form of a national or international award, and then they may perhaps gain access to the resources provided by the institutions of the national/regional innovation system.

To summarise, our interviews reveal great dissatisfaction with the activities of the supportive institutions and a significant lack of co-operation between independent inventors and institutions.[18].

DIFFERENCES BETWEEN MALE AND FEMALE INNOVATORS

Several studies argue that gender dynamics is also present in innovation. Throughout history, women have been denied access to certain cultural facilities and have not been given the chance to express their creative potential within their environment (Baker, 1998, 43). Experience shows that understanding innovation in terms of a technological product restricts the definition of innovation. This narrow definition relates to the specific position women have held within the socio-historical context that is, in a slightly modified and weaker form, still reflected in our times.

Given the above arguments, it may be expected that gender-related differences also exist when it comes to the formation of social capital and gaining access to the innovation support provided by the external environment. The definition of innovation may only be changed through the planned introduction of the female experience, something that has been overlooked throughout history. By experience, in the context of this chapter we mean the manner in which women gain information for developing and marketing their product and how they establish formal and informal connections within their environment and with the institutions they turn to for help and co-operation.

THE IDENTITY OF THE FEMALE INVENTOR

In her interview, one inventor who was given an award at the Seoul International Innovation Fair states that being an "inventor" is an atypical vocation, not unlike being a female engineer, and that she feels uncomfortable using the word "inventor." *"What is this strange word inventor? They asked me whether they could call me an inventor and I replied I was a designer; inventor is not the correct word; it is a little misplaced, like engineer."* (R 16)

Our interviewees note that only a small share of women engage in invention activities. They blame this on the female "routinised mind," namely, their disinterest in anything beyond family life and work commitments. In summary of the findings to date, and in reference to the theoretical framework, we can highlight the following reasons for such a low involvement of women:

- Innovation is an "atypical" profession for women. Innovation is narrowly defined and is usually related to technological activities.
- Women are burdened by household chores and caring for their children. The unequal distribution of household chores between the genders inhibits female creativity and limits it to only a few fields.
- Educational and other institutions reaffirm stereotypes (which are present at all levels of social life), which is apparent in the forms of vocational segregation presented above.

In the Western paradigm, the fixation of the social role defined by gender and grounded in biology is so strong that the social aspects of biology are understood (interpreted) only through a two-part model made up of the opposition of masculinity and femininity within the scope of characteristics and tasks pertaining to each model (including the division of family, vocational, and household tasks). Our female respondents highlight the fact that throughout history men have had the privileged position of "creator," namely, as the gender that generates ideas. As one female inventor put it, her male colleagues always find the time and space for creativity—for example, in the afternoon after work, while women have no such time and/or cannot make the time as they have to take care of their children. *"Men have more time . . . if women had more time, more quality time . . . women often have to engage in . . . they have their own job, and then they have work at home, and simply run out of time and energy to think about other things."* (R 17)

In modern times, we are faced with the need to re-assess the term "innovation," as well as the need to find new definitions to include female activities. Here we discuss the way in which women who have already entered this field (namely, women who have already patented their own products or are members of societies and similar organisations dealing with this field) generate their own social networks through which they obtain information on gaining funds, product marketing, and how they connect to other male and female creators.

NETWORK CREATION—ENCOURAGEMENT FROM THE INFORMAL ENVIRONMENT

In the previous sections we illustrated the fact that, compared to older ones, young inventors are more likely to perceive their family as a private network contributing to their innovative activities. Looking at the gender differences (middle-aged) female inventors often mention their family when recalling their creativity, which reflects the fact that they are still, to a greater degree than men, tied to the family, despite the fact that encouragement from the family is not received only positively. Namely, this "dual perspective" can be seen in the female respondents' answers. As one inventor put it, *"In some respects yes, in some respects no; when I attempted to become self-employed, my family was against it and I experienced great difficulties becoming independent."* (R 16)

Other respondents (like the one above) emphasise that people from their environment and their family members, who are often engaged in other activities, fail to understand their work. According to one interviewee, *"Although my immediate environment supports me, my family does not, and I don't feel understood . . . I am best understood by people similar to myself."* (R 17) Or, in the words of another interviewee, *"With respect to support . . . of individuals . . . in my town, there are one or two such people . . . in other respects, perhaps my sister, but only on a personal level, as she is not interested in such matters."* (R 16)

As is the case with their older male counterparts who we found needed no "non-professional" contacts, we also found that the women's contacts are chiefly limited to a few select people they meet as part of their work. Like their older male counterparts, middle-aged women (and partly also younger women) mainly seek support and encouragement from individuals with whom they can share their professional experiences. Furthermore, we found that women are to a greater extent than their male colleagues involved in family life and communication with family members. Women's involvement in family life can be compared to that of younger men who, in contrast to their older male counterparts, more often note a positive relationship between their family and their creativity. As a private network, the family supports their creativity; at the same time, they are not characterised by the "dual perspective" which characterises the female evaluation of this relationship.

ESTABLISHING BROADER PROFESSIONAL TIES AND SECURING INSTITUTIONAL SUPPORT—NEGATIVE DYNAMICS

In establishing broader professional ties, for example, with institutions and organisations that could provide financial and/or promotional support for their inventive ideas (and product launching), women are mainly restricted to co-operation with associations, most often the association of Slovenian inventors (ASI). *"However, I must say that the ASI is very active, you can make agreements through them; however, all other organisational and marketing services must be paid for."* (R 17)

Women often mention the advantages of "personal" communication which is not simply limited to telephone conversations and formal meetings. Such communication is only possible through associations (in most instances the ASI is mentioned) where relationships are not strictly formalised, while representatives of the associations are not "profiteers" only interested in doing business, but mainly enthusiasts who are prepared to help and provide consulting as part of the broader material perspectives defined by the concept of the association. In the words of one respondent, *"Yes, the ASI is a more personal association . . . for me it is very important to talk to a person and not an institution . . . Only personal contact means something."* (R 16)

Since the association in question—the inventors' association ASI—cannot support women in their innovation efforts, they are generally forced to finance their products themselves. As one female respondent put it, *"I invested my own funds; I had some savings from previous work, also some investments . . . Now I am looking for a grant, without success so far."* (R 17).

Our female respondents are not very successful in forging broader relationships with their external environment. They are also generally not very active in establishing broader contacts; that said, some represent the exception to the rule in their establishing of networks. As one female respondent emphasises, from the very beginning she established a well-branched network which she developed in different ways, *"I use the Internet a lot, I used it to search for companies, while some companies were also referred to me . . . I tried to attend most innovation events in Slovenia and I met a large number of people. I also made some new acquaintances at the Chamber of Commerce and Industry . . . and slowly I have come full circle."* (R 14)

She has also established contacts abroad, *"I spoke to Swiss and German inventors. There, the conditions are better . . . The countries have more wealth and they employ people to support inventors."* (R 14)

She emphasises that she had to do all the work from the idea through to promotion of the prototype and marketing the product by herself. Despite active involvement, she received support primarily from the ASI mainly because institutional support mechanisms are not fully developed or, in the words of the inventor, "there are none in Slovenia."

Institutional support can be best illustrated with a statement by an award-winning female inventor for whom the award partly paved the way for her professionally and creatively, *"The publicity came in handy after I was awarded the gold medal and as a result I was on TV. I was called by quite a number of people who wanted to work with me."* (R 16)

Like their older male counterparts, in some respects women expect to be supported by state institutions, but are at the same time distrustful and critical of them. Older men are unsuccessful in developing broader social networks and acquiring capital because of their lack of knowledge, for example, concerning use of the Internet. However, middle-aged and younger women do not display a lack of the knowledge that could be used to find the necessary information since many of them gained that knowledge in their workplace. The following two factors inhibit their establishing of broader professional contacts:

- Distrust of institutions and organisations where help could be obtained.
- A lack of family support: because of overburdening, the intertwining of family and professional life, women do not have the time to "establish broader networks." The only exceptions are female inventors who successfully combine their profession and inventive activities and obtain appropriate support from the institutions that employ them.

DISCUSSION AND CONCLUSION

In this chapter we focus on the importance of social capital for the innovative performance of independent inventors. By employing the case study method and using results of an analysis of interviews conducted with 22 independent inventors—entrepreneurs and individual inventors—in Slovenia[19] we investigated characteristics of the accumulation of social capital by independent inventors and the ways they collaborate with the external, innovation supportive environment at the national and regional/local levels. In doing so, we took account of the distinction between individual and structural social capital (Masciarelli, 2011). We argue that both dimensions are important when speaking of independent inventors.

Research findings so far indicate that a lack of opportunities to collaborate in teams and networks and thus to share and create knowledge—particularly tacit knowledge which can only be accessed through personal communications—negatively affects the quality of their innovations. Interactions among members of networks enable the creation of trust which is a precondition for knowledge sharing and genuine collaboration. The findings from this study suggest that the low level of trust (in people and institutions) characterising our respondents represents the main barrier to co-operation in teams and participation in networks, especially the open ones. The building of trust through interactions in networks appears especially important for independent inventors since they do not have either organisational social capital or organisational support available. An atmosphere of trust should thus contribute to the free exchange of knowledge between the network members as they should not feel that they have to protect themselves from others in the network.[20]

We may conclude from our findings that structural social capital is likely to be of the utmost importance in the second phase of the innovation process when independent inventors try to make their invention visible in the form of a patent/trademark and bring it to the market. To achieve this, they need a wide range of knowledge and skills as well as other resources like finance. Connections to a greater variety of networks would help them identify the availability of these resources and facilitate their utilisation.

The respective literature points to structural social capital in terms of connections to formal and non-formal networks outside primary groups as the type that actually enables access to more diverse knowledge and other resources. Our findings are in line with such assertions. They suggest that the structural social capital accumulated in national/regional innovation system institutions is more likely to facilitate access to supportive activities and resources that are usually provided to corporate inventors by their firms and institutes. However, it is likely that independent inventors lack both the opportunities and abilities needed to establish relations with diverse open networks, especially formal ones. They rely more on the social capital generated in closed networks (family, friends). This is not to suggest that social capital generated in closed networks is less important for independent inventors. Indeed, it may be less essential for highly complex specialist technical knowledge sharing and creation. Yet concluding from the information provided by our interviewees, it is very important in sharing tacit knowledge (informal learning, professional experiences). Moreover, factors such as a stable family environment may facilitate individual creativity and inventiveness. As discovered in our analysis, families (parents, spouses) may assume important assets like the time to innovate and—often—financial resources.

Our finding that civil society structures are the main source of both the individual and structural social capital of independent inventors is in line with the empirical evidence provided by Hauser et al. (2007) that 'associational activity' represents a robust influence on patenting activity. Yet, as noted by Westlund (2006), authors dealing with social capital in relation to innovation stress that social capital is also created in formal institutions—public and private. Having no access at all to institutional social capital or to just a limited amount of it, as suggested by our findings, means that independent inventors cannot take advantage of the resources these institutions provide. Since these are resources that could make up for the missing organisational support, independent inventors are further marginalised compared to their corporate colleagues.[21] Extending their ties to formal institutions and the networks established/facilitated by these institutions would thus considerably improve their invention resources. In line with our findings, the social capital created in closed networks (friends, schoolmates, etc.) may be very valuable for facilitating access to the social capital created in formal institutions. The same goes for civil society organisations and professional associations.

It was assumed that in the Internet era independent inventors' access to knowledge and other resources is less dependent on relations with various networks. However, it may be proposed that while ICT may be a valuable tool enabling access to information it is unlikely that participation in virtual networks could become a new way of collaboration facilitating independent inventors' innovation activities, irrespective of the age cohort.

One question addressed by this chapter considers gender differences in creating social capital. Our findings confirm that gender plays an important role in both access to social capital and external support. It appears that closed networks provide more support to male inventors than female inventors. Women are characterised by a "dual perspective," meaning that on the one hand women are generally more involved in family life than men and obtain certain support from this while, on the other hand, family dynamics appears to be a major obstacle to their creativity. However, in terms of establishing broader relationships both men and women are equally hindered by their distrust of institutions. Female inventors apparently depend more on their own resources and abilities to recognise and utilise the structural social capital created in open networks in both formal institutions and civil society networks.

Following Lin et al. (2001) who link the benefits of co-operating with actors in the (external) innovation environment with the power of individual members of collaboration networks, we could argue that independent inventors lack visibility and power. This affects their opportunities to collaborate in regional networks as well as the scope and type of benefits they can draw from such collaboration. This also applies to their access to benefits from spill-overs of innovative knowledge created in regional networks.

NOTES

1. A concept of the national innovation system was established at the end of 1980 (Freeman 1987; Lundvall, 1988). It is based on the premise that understanding the links between the actors involved in innovation is the key to improving technology performance.
2. There are marked differences in the relative role and weight of different institutions in the national innovation system among countries which partly account for the focus on the country level (OECD, 1997, 12). A number of framework policies relating to regulation, taxes, financing, competition, and intellectual property can facilitate or inhibit the various types of interactions and knowledge flows.
3. Hence, policy strategies could be oriented to the promotion of accessibility in the development of a regional innovation system (Andersson and Karlsson, 2002) and the development of local comparative advantages linked to specific local resources (Maillat & Kébir, 2001).
4. For example, in 2005 in Norway, 52% of patents were filed by individual inventors, in Ireland 43%, in Belgium 42%, in Austria 41%, in Finland 35%, in Sweden 34%, and in Italy 24% (*http://www.invention-ifia.ch/independent_inventors_statistics.htm*; accessed January 30, 2012). Accordingly, individual inventors, or more generally independent inventors, represent an important factor of national innovation systems.
5. Contextually connected information enables the development or improvement of products or manufacturing processes (Landry et al., 2002).
6. Knowledge-based innovation requires not one but many kinds of knowledge. The development of intangible assets is becoming crucial in strengthening learning capacities. To some extent, these assets can be seen as a specific form of capital that is derived from social relations, norms, values, and interaction within a community (Landry et al., 2002).
7. We also obtained information from focus group discussion conducted as part of the project. Seven representatives of independent inventors and supporting environment institutions participated in the discussion. A transcript is available from the Social Data Archive (*http://adp.fdv.uni-lj.si*; accessed January 30, 2012.)
8. Åstebro and Thompson (2007, 2) divide independent inventors into individual inventors and inventor-entrepreneurs. In most cases, inventor-entrepreneurs use their inventions themselves in their own companies, while individual inventors strive to license their inventions to other companies. However, this distinction is not always clear as some inventors assume different roles during their life: licensing their patents, establishing their own enterprises, and also being employed as consultants for other companies.
9. According to some studies, there are only 2% to 3% of female inventors (Mariani & Romanelli, 2006; Giuri et al., 2006).
10. The inventors in our sample are slightly older than in other studies researching independent inventors (Davis et al., 2009—51.4 years; Wagner, Weick, & Eakin, 2005—50.5 years).
11. Six inventors younger than 40 years of age were included in the sample of interviewed Slovenian inventors.
12. The cases of the younger inventors from our sample support the statement that social capital and human capital are transmitted from parents to children, thereby confirming older and contemporary studies (Coleman 1988, Kaasa, 2007).
13. The majority of the sample either has a tertiary or high school level of education and their parents have an education one level lower.

14. Which they select diligently as the costs of these, especially those abroad, are very high and out of reach of most of them.
15. In this regard, it is worth noting that uncertainty avoidance mechanisms differ with regard to social, regional, and cultural environments (Waarts & van Everdingen, 2005).
16. In Slovenia, entrepreneurs can apply for funds used to develop innovations through public calls such as grants for inventive start-up companies from the Slovenian Enterprise Fund or the "innovation voucher" organised by the Public Agency of the Republic of Slovenia for Entrepreneurship and Foreign Investments. The latter covers up to 50% of costs that lead to the patent application, industrial design, or trademark of a new product. Although individual inventors cannot apply for these public tenders, at most supporting organisations they can at least obtain free advice concerning their inventive projects, especially on how to set up their own enterprise.
17. At the time the study was conducted, the chosen sample only included a few inventors with start-ups so we can assume that the majority of inventors have had no true experience with such organisations even though they expressed very negative opinions of them.
18. Existing institutions in Slovenia, partly with the exception of the inventors' association, do not serve the needs of independent inventors, especially individual inventors, who do not want to establish their own enterprises. Institutions like the Foundation for Finnish Inventions that financially support the early phase inventive activities (see Georghiou et al., 2003, 146), or a technology transfer office that would help independent inventors license their patents or a prototype workshop financed from state sources, do not exist in Slovenia.
19. We remind the reader that the methodology applied does not allow the generalisation of the findings. They are limited to the sample of the interviewed independent inventors.
20. Indeed, Hauser et al. (2007) provide empirical evidence at the national level that it is the connectedness of people which has a major impact on innovation in [industrialised] countries, while trust may have a more robust impact on economic growth at a national level.
21. On one hand, independent inventors exhibit a high degree of distrust in institutions established by the state and private initiatives which, however, is not always justified and might also be the result of a lack of certain skills and competencies on the part of independent inventors. On the other hand, it is true that the national innovation system, including the innovation support agents, marginalises independent inventors and systematically overlooks their specific needs.

REFERENCES

Adam, F., Hafner, A., Podmenik, D., Podmenik, D., Rončević, B., Šinkovec, U., & Vojvodić, A. (2010) *Inovativna jedra v regionalnem razvoju*. Ljubljana: Vega.

Anderson, H. (2004) Why big companies can't invent. *Technology Review*, 107, (4), 56–59.

Andersson, M., & Karlsson, C. (2002) Regional innovation systems in small and medium-sized regions. A critical review and assessment. Working Paper 2002–2. Jönköping: International Business School. http://www.ulb.ac.be/soco/asrdlf/documents/RIS_Doloreux-Parto_000.pdf. Accessed March 1, 2012.

Åstebro, T., & Thompson, P. (2007) Entrepreneurs: Jacks of all trades or hobos? http://www.lanzapymes.com/DOCUMENTOS_G/DESCARGAS/ENTREPRENEURS%20

JACKS%20OF%20ALL%20TRADES%20OR%20HOBOS.PDF. Accessed May 9, 2011.
Baker, B. (1998) *Philosophy and sex*. New York: Prometheus Books.
Bourdieu, B. (1986) The forms of capital. In Richardson, J. (ed.), *Handbook of theory and research for the sociology of education*. New York: Greenwood, pp. 241–258.
Brown J. S., & Duguid, P. (2001) Knowledge and organization: A social-practice perspective. *Organization Science*, 12, 198–213.
Coleman, J.S. (1988) 'Social capital and the creation of human capital. *American Journal of Sociology* (Supplement 'Organizations and Institutions'), 94, S95–S120.
Cooke, P. (2001) Regional innovation systems , clusters and the knowledge economy . *Industrial and Corporate Change*, 10, 945–974. http://webappo.web.sh.se/C1256CFE004C57BB/0/3B57D93FDE3D6405C125768000476506/$file/Regional%20systems%20of%20innovation.pdf. Accessed November 21, 2011.
Cooke, P. (2005) Regionally asymmetric knowledge capabilities and open innovation: Exploring 'Globalisation 2'—a new model of industrial organisation. *Research Policy*, 34, 1128–1149.
Cooke, P., Boekholt, P., & Todtling, F. (2000) *The governance of innovation in Europe*. London: Pinter.
Cooke, P. (2005) Regionally asymmetric knowledge capabilities and open innovation: Exploring 'Globalisation 2'—a new model of industrial organisation. *Research Policy*, 34, 1128–1149.
Dahlin, K., Taylor, M., & Fichman, M. (2004) Today's Edisons or weekend hobbyists: Technical merit and success of inventions by independent inventors. *Research Policy*, Elsevier, 33 (8), 1167–1183.
Dakhli, M., & De Clercq, D. (2004) Human capital, social capital and innovation: A multi-country study. *Entrepreneurship & Regional Development*, March 16, 107–128.
Danilda, I., & Granat Thorslund, J. (eds.). (2011) *Innovation and gender. http://www.vinnova.se/upload/EPiStorePDF/vi-11-03.pdf. Accessed July 2, 2012*.
Davis, L.N., Davis, J., & Hoisl, K. (2009) What inspires leisure time invention? http://www2.druid.dk/conferences/viewpaper.php?id=5578&cf=32/ Accessed June 9, 2011.
De Filippis, J. (2001) The myth of social capital in community development. *Housing Policy Debate*, 12, 781–806.
Densin, N.K., & Lincoln, Y.S. (eds.). (1994). . *Handbook of qualitative research*. Thousand Oaks, CA: Sage.
EC. (2011) Innovation *Union competitiveness report*, 2011 edition. Luxembourg: Publications Office of European Union. http://ec.europa.eu/research/innovation-union/pdf/competitiveness-report/2011/iuc2011-full-report.pdf#view=fit&pagemode=none. Accessed October 24, 2011.
Edquist, C. (2004) Systems of innovation—A critical review of the State of the Art. In Fageberg, J., Mowery, D., & Nelson, R. (eds.), *Handbook of innovation*. Oxford: Oxford University Press.
Fleming, L. (2006).'Lone inventors as the source of technological breakthroughs: Myth or reality? http://www.hbs.edu/units/tom/docs/lfleming.pdf. Accessed February 29, 2012.
Fleming, L. (2007) Breakthroughs and the 'long tail' of innovation. *MIT Sloan Management Review*, 49 (1), 69–74.
Freel, M.S., & Harrison, R. (2006) Innovation and cooperation in the small firm sector. . *Regional Studies*, 40 (4), 289–305.

Freeman, C. (1987) Technology, policy and economic performance: Lessons from Japan. London: Pinter.
Fuglsang, L. (2008) Innovation and the creative process: Towards innovation with care. Cheltenham: Edward Elgar Publishing.
Georghiou, L., Smith, K., Toivanen, O., & Ylä-Anttila, P. (2003) Evaluation of the Finnish Innovation Support System. Helsinki: Ministry of Trade and Industry Finland.
Giuri, P., et al. (2006) Everything you always wanted to know about inventors (but never asked): Evidence from the PatVal-EU survey. http://cosmic.rrz.uni-hamburg.de/webcat/hwwa/edok07/f10970g/dp06-11.pdf. Accessed August 31, 2011.
Guiso, L., Sapienza, P., & Zingales, L. (2004) The role of social capital in financial development. *American Economic Review*, 94 (3), 526–556.
Giuri, P., Et al. (2006) Everything you always wanted to know about inventors (but never asked): Evidence from the PatVal-EU survey. *http://cosmic.rrz.uni-hamburg.de/webcat/hwwa/edok07/f10970g/dp06-11.pdf*. Accessed August 31, 2011. (31.8.2011).
Hauser, C., Tappeiner, W., & Walde, J. (2007) The learning region: The impact of social capital and weak ties on innovation. *Regional Studies*, 41 (1), 75–88.
Kaasa, A. (2007) Effects Of different dimensions of social capital on innovation: Evidence from Europe at the regional level. University of Tartu, Faculty of Economics and Business Administration, Working Paper Series, No. 51. Tartu: University of Tartu. *http://ideas.repec.org/s/mtk/febawb.html*. Accessed September 9, 2011.
Kaasa, A., & Vadi, M. (2008) How does culture contribute to innovation? Evidences from European countries. Tartu: Tartu University Press. *http://www.mtk.ut.ee/orb.aw/class=file/action=preview/id=423461/febawb63.pdf*. Accessed March 3, 2011.
Knack, S., & Keefer, P. (1997) Does social capital have an economic payoff? A cross-country investigation. *Quarterly Journal of Economics*, November, 1251–1288.
Landry, R., Amara, N., & Lamari, M. (2002. Does social capital determine innovation? To what extent? *Technological Forecasting and Social Change*, 69 (7), 681–701.
Lettl, C., Rost, K., & von Wartburg, I. (2009) Why are same Independent Inventors 'heroes' and others 'hobbyists'? The moderating role of technological diversity and specialisation. *Research Policy*, 38 (2), 243–254.
Lin, N., Cook, K.S., & Burt, R.S. (2001. *Social capital. Theory and research*. New York: Aldine de Gruyter.
Lundvall, B.A. (1988) Innovation as an interactive process: From user-producer interaction to the national system of innovation. In Gossi G., Freeman, C., Nelson R., Silverberg, G., & Soete, L., *Technical change and economic theory*. London: Pinter.
Maillat, D., &and Kébir, L. (2001) Conditions-cadres et competitivite des regions: une relecture. *Canadian Journal of Regional Science*, 24 (19), 41–56.
Mariani, M., & Romanelli M. (2006) "Stacking" or "picking" patents? The inventors' choice between quantity and quality'. *http://www.dime-eu.org/files/active/0/MarianiRomanelli.pdf*. Accessed August 31, 2011.
Masciarelli, F. (2011) *The strategic value of social capital. How firms capitalize on social assets*. Cheltenham: Edward Elgar Publishing.
Nicholas, T. (2011) Independent inventors during the rise of the corporate economy in Britain and Japan. *Economic History Review*, 1–28. http://people.hbs.edu/tnicholas/IndepBR_JP.pdf. Accessed September 10, 2011.

OECD, (1997) National Innovation Systems. Paris: Organization for Economic Cooperation and Development. *http://www.oecd.org/dataoecd/35/56/2101733. pdf. Accessed September 19, 2011.*
Paxton, P. (1999) Is social capital declining in the United States? A multiple indicator assessment. *American Journal of Sociology,* 105, 88–127.
Perry-Smith, J.A., & Shalley, Ch.E. (2003). The social site of creativity: A static and dynamic social network perspective. *Academy of Management Review,* 28, 89–106.
Plankenhorn, S. (2008) *Innovation offshoring.* Wiesbaden: Gabler.
Podmenik, D., & Hafner, A. (2011) Case study: Independent inventors. In Adam, F., & Westlund, H. (eds.), *Socio-cultural dimensions of innovation performance.* Ljubljana: IRSA, 131–155.
Powell, W.W., Koput, K.W., & Smith-Doerr, L. (1996) Interorganizational collaboration and the locus of innovation: Networks of learning in biotechnology. *Administrative Science Quarterly,* 41, 116–145.
Putnam, R.D. (1993) *Making the democracy work. Civil tradition in modern Italy.* Princeton: Princeton University Press.
Siisiäinen, M. (2000) Two concepts of social capital: Bourdieu vs. Putnam. Paper presented at ISTR Fourth International Conference "The Third Sector: For what and for whom?" Trinity College, Dublin, Ireland, July 5–8, 2000. http://dlc.dlib.indiana.edu/dlc/bitstream/handle/10535/7661/siisiainen.pdf?sequence=1. Accessed February 6, 2012.
Singh, J., & Fleming, L. (2010) Lone inventors as the source of technological breakthroughs: Myth or reality? *Management Science,* 56 (1), 41–56.
Von Hippel, E. (2007) Horizontal innovation networks—by and for users. *Industrial and Corporate Change,* 16 (2), 1–23.
Waarts, E., & van Everdingen, Y. (2005) 'he influence of national culture on the adoption status of innovations: An empirical study of firms across Europe. *European Management Journal,* 23, 601–610.
Wagner Weick, C., & Eakin C.F. (2005) Independent inventors and innovation: An empirical study. *http://www.inventored.org/papers/Weick.pdf. Accessed April 4, 2011.*
Westlund H. (2006) *Social capital in the knowledge economy: Theory and empirics.* Berlin, Heidelberg, New York: Springer.
Westlund, H. (2011) Collaboration in innovation systems and the role of social capital. In Adam, F., & Westlund, H. (eds.), *Sociocultural dimensions of innovation performance.* Ljubljana: IRSA, pp. 108–130.
Whalley, P. (1992) Survey of independent inventors: An overview. *http://www.osti.gov/bridge/servlets/purl/10187932-DeF7BU/10187932.pdf. Accessed February 6, 2012.*
Wolfe, D. (2002) Knowledge, learning and social capital in Ontarip's ICT clusters. Paper prepared for the annual meeting of the Canadian Political Science Association, University of Toronto. Toronto, Ontario, May, 29–31.
Wuchty, S., Jone, B.F., & Uzzi, B. (2007) The increasing dominance of teams in production of knowledge. *Scienceexpress,* 316, 1036–1039.

9 "Individuals" Networks and Regional Renewal
A Case Study of Social Dynamics and Innovation in Twente

Paul Benneworth and Roel Rutten

INTRODUCTION

It is now widely accepted that economic growth and wealth generation are dependent upon innovation, in turn dependent in part on human capital—the knowledge embedded within the people who innovate. Effective innovation depends on individuals' capacities to create, circulate, combine, and exploit knowledge capital from a range of sources to create new products, processes, or techniques. This relationship of innovation and knowledge capital faces a perennial problem because some of the most important forms of knowledge capital are in effect 'tacit,' that is to say, difficult to codify and transmit. Handling that knowledge is in part performative, coming about through the activities, networks, and contacts in which individuals are involved.

There is an emerging appreciation of the fact that knowledge transfer and exchange involving knowledge transfer is at least partly a socialised and collective process (Lagendijk, 2007). People engaged in similar kinds of knowledge creation and innovation activities cohere into communities that have a partly social dimension which facilitates cognitive proximity (Boschma, 2005); this might be in local work settings, e.g., Communities of Practice (Wenger, 1998), in more distributed and remote communities, (cf. 'networks of practise,' Benner, 2003), or more imagined communities, what Haas (1992) referred to as 'epistemic communities' who share norms and conventions despite only a loose coupling between community members. Understanding the dynamics of innovation and its economic growth consequences necessitates better understanding those social processes, not only their economic manifestations and functions but also their wider social dynamics.

One dimension on which social communities differ is that of territory, and there is an intuitive connection between this turn toward more sociological understanding of innovation, and existing territorial approaches, which consider how place-specific hard institutions and soft cultures shape territorial innovative capacities. In this chapter, we are concerned with the family of what Moulaert & Sekia (2003) call 'territorial innovation models,' the family of cognate disciplines concerned with understanding the spatial dimension of territory and growth. There are a number of branches of the TIM family which have a specific social dimension. One example is in 'cluster' approaches to territorial innovation, where Gordon and McCann (2000) discerned that one branch of cluster theories was based on social capital explanations (the other two being based upon urbanisation economies and input-output linkages).

The treatment of culture within TIMs is characterised by a tendency to instrumentalise culture as an explanation for unique situations that cannot otherwise be explained (Oosterlynck, 2007). In this chapter, we seek to contribute to the debates more generally in this volume and elsewhere, the aim of which is to better specifying precisely how social relationships and 'culture' influence firm innovation, and hence economic development. But this cultural dimension has never been satisfactorily analysed: 'good' regions have by definition learning cultures, which robs the concept of its potentially explanatory powers. So we seek to critically analyse the concept both theoretically and empirically to gain an insight into how social dynamics may better be incorporated into territorial innovation models. Our argument is that there is a need to distinguish the territorial effects on individual innovators' ego networks with many different kinds of connections, and on the social networks (e.g., clusters) which exist specifically for the purpose of supporting innovation. Better understanding the separate territoriality of these two classes allows a much sharper distinction to be drawn in answering the question of how territorial is territorial innovation, and to support Sheamur's wider call to wider space as an opportunity field rather than as hard lines limiting interactions.

In this chapter, we address the research question of how do the social habits, norms, and routines of local innovation networks shape the wider culture and behaviour of their particular local territory and the local innovation environment as well as economic development outcomes. To address this question, we begin by looking more closely at TIM concepts, and highlight the continuing critique of their failure to analytically specify social variables and dynamics as more than an emergent function of success. We then turn to explore a case study of a region in which there has been a shift in its learning capacity, which while not representing a totalising cultural shift nevertheless represents a qualitative change in its culture associated with—but not entirely driven by—a qualitative change in its innovative capacity. We then turn to analyse this process of cultural change, and highlight in particular the way that social dynamics at a variety of scales—the individual, the network, the regional-political, and the regional-cultural—interact as part of this change. These different scales are linked by hinging activities which bring—for example—individuals together into networks. What TIM exponents consider to be change in the regional culture is built upon parallel but separate changes in individual's networks as well as innovation transaction networks. We therefore conclude by arguing that a more comprehensive approach to studying regional learning cultures therefore needs to make precisely this distinction a great deal more clearly to help bring rigour back to the study deal more rigorously with this inter-relation between these scales at which innovation takes place.

INNOVATION DYNAMICS AND THE REGIONAL CULTURAL CONUNDRUM

The Rise and Fall of Territorial Innovation Models (TIMs)

A range of recent trends all pointed to an economic geographical shift away from the manufacturing economy to a knowledge-based economy with a highly uneven geography very different from that of the industrial economy. Moulaert and Sekia (2003) identified what they called the Territorial Innovation Model (TIM) family of concepts which attempted to make sense of this wider shift. TIMs had emerged in parallel in different disciplines interested in explaining differential economic performance in terms of different territorial innovation performance. Their argument was that all the models were in some way responding to the new reality that increasing economic globalisation had placed a premium on local capacities as a determinant of those local outcomes. Uniquely among contemporary economic activities, innovation was dependent on face-to-face contact to transmit the necessary tacit knowledge, and that contact in turn depended on the degree of proximity present (Boschma, 2005). As a result, many cognate disciplines had developed their own explanatory theories for that localisation process. Moulaert and Sekia argued for five basic models; local production systems; innovative milieus; learning regions; regional innovation systems; new industrial spaces; and spatial clusters of innovation. Lagendijk (2006) argued that these could be better understood as a cognate genealogy, with successive generations, structuralist-organisational, institutional-conventional, and cognitive. Each model was rooted in its own disciplinary history, with its own conceptual clarity, intellectual integrity, and clarity both about the conditions to which it is applicable, as well as the extent of findings which can be inferred from its application.

These explanations have not been uncontentious, and have been subject to critique from a variety of perspectives.[1] It is possible to distinguish two critiques operating at two scales, first within the trajectories of individual models and second with the ideas of TIMs as a whole. In terms of individual models, the very brief examples of 'clusters' was a TIM which was comprehensively attacked for being a slippery concept which elided between theoretical foundations at the whim of policy-makers (Gordon & McCann, 2000; Martin & Sunley, 2003; Benneworth & Henry, 2004; Karlsson, 2008). Second, critiques are leveled at TIMs more generally, and highlight four particular problems (cf. Christopherson & Clark, 2010) recurrent in these issues, although not necessarily applicable to all TIMs.

The first of these critiques is that there has been an implicit assumption that knowledge assets are somehow intrinsically 'regional' (Gertler & Wolfe, 2006). One manifestation has been an uncritical focus on politically bounded spaces rather than looking at the wider networks within which innovation takes place (what Cooke, 2005, calls the 'spatial envelope' into which the region is placed). Second, it overlooks the importance of non-regional knowledge and the fact that knowledge brought into the region through, for example, foreign investors, has its own topologies of power its regional application (Kotzschatzky, 2009). While Storper (1995) points to untraded interdependencies and knowledge spillovers as drivers of industrial districts, Christopherson and Clark (2007) provide compelling evidence of how lead companies in one particular industrial district in Northeast America deliberately limit the free flow of knowledge in the supply chain as a competitive strategy. Japanese car makers follow a similar strategy in the Indonesian automotive industry, where the development of an indigenous innovation potential is hampered by Japanese efforts to prevent knowledge from flowing to their supplier network (Irawati & Rutten, 2011).

A second critique is that the ideas are too specific to a particular kind of industrial region in North America and Europe, particularly those involved in high-technology manufacturing. Part of this relates to the genesis of the ideas which can be traced to understanding a wider socio-economic shift from Fordist manufacturing to a post-Fordist, post-industrial form of economic organisation (cf. Piore & Sabel, 1984; Albrechts & Swyngedouw, 1989). Also within this can be identified the role of policy-makers who have sought solutions for their regions with problems, which has encouraged the uncritical transfer of policy measures between regions. This has the effect of making these concepts appear in regions where they are not necessarily applicable (Hassink, 1993; Lagendijk & Cornford, 2000; Hassink & Lagendijk, 2001).

A third critique is that the TIMs are undertheorised and that there has been conceptual borrowing and elision between the TIMs, which has left them as 'small' theories subject to but incapable of challenging external global 'big' powers and structures (Amin & Palan, 2001). This argument has most effectively been developed by Markusen (1999) who critiqued many studies in practice for failing to really test theory via empirics, instead defaulting toward merely selectively illustrating theoretical contentions. Lagendijk (2003) argued that there was a rush toward the careless development of concepts which were never rigorously empirically tested and which became the basis for further theoretical developments. Hudson's (2003) argument was that this had served to detach the concepts from wider political economies of power and reify the idea of the 'local' as subordinated to the 'global,' which in turn served a particular kind of neo-liberal economic development agenda.

The final critique was that the success of TIMs had been driven by policy-makers who had funded academics to give their normative ideas a veneer of scientific plausibility, or in a less sceptical view of the process, had encouraged academics down a particular theoretical path which was appealing to them as policy-makers (Lovering, 1999; Martin & Sunley, 2003; Lagendijk, 2007). Certainly, it is hard to refute the accusation that a huge number of empirical studies were undertaken which had very little connection to theory, undermining serious comparisons and further detaching theoretical developments from empirical efforts (Coenen, 2006). But the criticism here was not so much one of rigour as of the negative effects of policy-makers rather than academics shaping the research agenda, with a sense that academics were required to find the 'right' results.

We acknowledge that all these critiques are in large measure justified and require a critical scepticism about the further validity of territorial innovation models that do not overcome these shortcomings. At the same time we make two further observations: first, these critiques do not undermine the fact that territorial innovation is an important determinant of economic growth, but what is needed are rigorous theoretical models which explain how uneven spatial development emerges. Second, in seeking to make TIMs rigorous, one can only deal with a single TIM (cf. Benneworth & Henry, 2005), and therefore we use a single TIM, namely that of the learning region.

Re-Specifying the Learning Region as a Rigorous Concept

The idea of the learning region emerged in a 1993 Storper article which sought to understand the competitive success of a limited number of industrial districts in terms building local connections between the global networks within which key actors were embedded. The idea became successful because of its wider policy and political support, fitting with policy-makers' interest, but also because it was a 'first-cut' way of describing something clearly becoming interesting to policy-makers and academics. The value of learning regions can be distilled down to understanding the process of learning in regional networks, places where actors interacted in ways that constructively stimulated mutual learning. In particular, this draws attention to the importance of social networks (cf. inter alia Amin & Cohendet, 2004; Morgan 2004; Moulaert & Nusbaum, 2005; Rutten & Gelissen, 2010). The problem with the concept of the learning region, identified as early as 1999 by Markusen, is that it is a fuzzy concept, and is a way of talking about something that scientists and policy-makers *feel* is important without acquiring the necessary academic rigour to be considered a concept (cf. Hassink, 2007, 2010; Christopherson & Clark, 2010). In this chapter, we suggest that this can be accomplished by seeking to decompose its various elements and identify what allows us to talk about what matters for regional growth and at the same time allow suitable rigorous treatment.

At the heart of the fuzziness lies a simple elision of actors and their regions afflicting all TIMs, but more problematic in the case of learning regions because of the assumption that the region becomes an agent (Markusen, 1999; Nuur et al., 2009; Uyarra, 2011). We concur with Sheamur (2011) in that the various problems identified with the learning regional concept and the TIM literature in general all relate to a more fundamental problem of a simple elision between actors and their regions. The idea of a learning region has become—for what are primarily strategic reasons associated with the way it has been used in practice *(cf.* Lagendijk, 2007)—strongly identified with a reading of regions in which regions are themselves actors with characteristics and agency, mediated through soft cultures and hard institutions (Trigilia, 1991; Morgan, 1997). This has created a conceptual difficulty in specifying how real actors are influenced by institutions and culture (Granovetter, 1985; Harrison, 1992). We highlight two assumptions between which fuzziness arises, namely, soft structuration and stickiness. In 'soft structuration,', there is the assumption that outcomes are produced in a dialectic tension between institutions and individuals: but critically this fails to address the problem of fuzziness. In stickiness, there is an assumption that individuals have a strong local preference that guides them to be able to disproportionately benefit from local innovation assets. In these readings, regions are things to which actors have a stickiness, and containers in which structuration processes occur.

Our argument is that this problem is effectively one of scale, and attempting to create narratives of meso-scale actors (regions) from micro-scale activities (innovation networks), which resorts to reifying cultural constructs as explanatory variables which can never be satisfactorily operationalized or tested (cf. Buesa et al., 2010). The reality of what is referred to by 'regions' in the context of TIMs is a space at which regular and repeated interactions take place (Lagendijk & Oïnas, 2005) rather a 'scale' that is above micro-actors. This definition sees the region as an emergence space for tacit knowledge exchange and socialised learning facilitating by inter alia physical proximity as well as social interactions (the cognitive proximity identified by Boschma, 2005, & Lagendijk, 2006). The focus must therefore shift to understanding how innovating actors interact and how in these interactions new cultures supportive of innovation emerge in those regions, and how these networks, and those cultures, support other actors in the innovation process. As a first step in this process, we ask three research questions to potentially identify promising future directions for conceptual development around relationships between innovation performance and the cultures of local innovation networks:

- What are the innovation resources which individuals draw upon in innovation activities which are accessed through social networks?
- How do the norms and dynamics of these innovation networks relate to wider local cultures? and

- How does the institutionalisation of loose networks and norms into wider social cultures and mores take place?

OVERVIEW OF CASE STUDY AND METHOD

To explore those questions, we use a case study where claims have been made for a shift toward an 'innovative culture,', Twente, in the east of the Netherlands, which for a time was the pre-eminent Dutch textiles region, earning it the name of the "Manchester of the Netherlands." From 1945, the region declined steadily, and entered a unemployment crisis in early 1970s, as firms became increasingly uncompetitive. The Regional Development Agency for the Province of Overijssel (*Overijsselse Ontwikkelings Maatschappij*, or OOM), was created in 1976 to reverse these problems, and this was apparently successful. By the mid-2000s, the National Spatial Development Strategy highlighted Twente as a region with significant innovative growth potential, and in 2010, a Consultancy report for the Dutch government argued that around the university was one of only three plausible high-technology clusters in the Netherlands (Buck Consulting, 2010).

The case study raises a more general question about the over-specification of territorial innovation models. Despite there being more learning

Figure 9.1 The location of the region of Twente in the Netherlands and Europe.
Source: *ITC, 2005 (Courtesy of Faculty ITC, Univ. Twente).*

192 *Paul Benneworth and Roel Rutten*

activities and successful innovation networks in Twente, claiming that the region is a 'Learning Region' clearly does not make sense: despite an innovative culture, the region still performs poorly with respect to other regions in Netherlands in terms of knowledge economy indicators such as qualification level wages, and unemployment. The region continues to suffer 'brain drain' where qualified graduates from its universities are forced to migrate to find suitable employment (Bijleveld & Geerdink, 2010). In this chapter, we seek to explore the dynamics of 'regional innovation cultures' and in particular the kinds of processes by which cultures influence business innovation, to flesh out tensions and hence identify potentially fruitful avenues for understanding innovation dynamics in their socio-cultural context.

This points toward the use of a case study approach using a thick description of a single case to gain a detailed structured insight into the research question under consideration to consider how these tensions play out in practice in long-term perspective. This of course provides a limitation on the wider theoretical claims that the chapter can make, and the main one of those is in terms of the comprehensiveness of the argument. Because we are working from a single case study, we cannot definitively establish the full breadth of the kinds of domains in which socio-cultural variables might influence innovation dynamics. But we can take a step forward beyond the problematic "Learning Region" model—to suggest some of what must necessarily be considered to deal with critiques of the approach.

The case study that we are using here is a synthesis of a number of separate pieces of research which have all attempted in some way to unpack Twente's regional paradox between evident regional successes and persistent regional problems. This research has involved around 80 interviews including a week-long peer review visit to the region as part of the OECD Universities and Regional Development project. Further detail on these interviews is provided in Table 9.1. below: this is at best able to give a flavour of interviews, because people have evolved over the course of the interviews, and been interviewed repeatedly in the course of the project.

Table 9.1 The Interviews within the Case Study by Time and Sector

Sector of interviewee		Time period (project phase)	2007
University manager	8	Summer 2004	39
Academic	7	Winter 2005	8
Policy organisations	5	Spring/Summer 2006	6
Support organisation	11	Summer 2007	9
Spin-off firm	21	Summer/Autumn 09	6
University services	16	2010	4
R&D intensive firms/ institutes	4		
Total	72	Total	72

The table shows interview events, by date and sector of the interview *at the time of the interview event.*

Some academic material has been used, including Clark (1998), Klein Woolthuis (1999), Lazzeretti and Tavoletti (2005) as well as policy material: this material is comprehensively listed in Benneworth and Hospers (2007) and not reproduced here, as well as Stuurgroep Kennispark (2008) and Bijleveld and Geerdink (2010). Material has been combined through a triangulation process where multiple sources are sought to establish key events, changes in regional direction, and the causes of those changes. That single narrative provides in turn a means to reflect on what have been the region learning processes, their connection to innovation, and in what sense there can be said to have been a change toward a more innovative and entrepreneurial 'regional culture.'

The cultures and activities described in this case study are still active and indeed are developing to this day. But in order to preserve the historical nature of the analysis, we conclude the historical period with which we are concerned in 2004 for two reasons. The first of these is practical; the research on Twente we draw on began in 2004, and taking 2004 as our end-point means that the case study is entirely historical. Second, it was in 2004 that the Dutch Ministry of Economic Affairs published its regional economic development strategy for the Netherlands, Peaks in the Delta (*Pieken in de Delta*), in which the Twente region (as part of the east of the Netherlands) had its role acknowledged nationally as the high-technology industrial region. Third, in being a 25-year case study, it is long enough to encompass a complete generational shift while still retaining a strong primary evidential record.

THE TWENTE REGION IN THE 1970S

The reason for starting the analysis in 1975 was that this was the date of the recognition of the depth of the crisis with the formation of the Provincial RDA, the OOM, and it is necessary to provide a little background context about the situation in which the region then found itself. The textile industry in Twente was created with a great deal of support from the *Nederlandsche Handelsmaatschappij* (NHM), created in 1824 as part of national reconstruction following the Napoleonic occupation. The Twente textiles industry was to process cotton grown in the Dutch East Indies colonies (today Indonesia). Twente's textiles industry emerged in an environment protected by national subsidies, a captive buyer of raw cotton imports from Dutch eastern possessions, and able to export these textiles back to the East behind protective tariff barriers (Klaassen, 1968). This allowed the region to flourish in textiles and support mechanical industries in the late 19th and early 20th centuries, with factory owners sharing their profits with the workforce, investing in, for example, a network of parks for their workers' relaxation, and offering rising living standards. Avoiding industrial unrest came at the expense of investment in higher productivity and quality, and lowering costs.

When the protected environment disappeared after World War II, with Indonesian independence, lowered tariffs, and the rise of low-wage economies, Twente's textiles industry entered a 40-year period of decline and almost total disappearance. The textile entrepreneurs were aware of the problems they faced, but under sudden competitive pressure faced the almost impossible task of merging to build critical mass, reducing employment, and creating new innovative products. There were serious efforts among the textiles entrepreneurs to secure their survival, and education played an important role in these efforts. A group of textile firm owners created the Academy for Art and Industry (Academie voor Kunst en Industrie or AKI) to help create a new class of designers to create novel mass-market products. A foundation was created in 1950 to lobby the national government for the creation of a Technical University in the northern and eastern provinces of Netherlands, and in 1964, the Twente Technical University (Technische Hogeschool Twente, or THT) was created to support regional business by creating a new innovative engineering class.

Other efforts involve mergers to create stronger businesses: the one textiles firm to survive in anything like its historical incarnation was Ten-Cate, formed from a merger between two textiles businesses in the west of Twente in 1957 (de Vries, 2005). The Royal Dutch United Textiles (KNTU) was created in 1962, and a series of mergers took place around Van Heek textiles, but these changes were not sufficient to allow adaptation to the new, cut-throat global textiles market. Over the period 1950 to 1980, there was a steady loss equating to around 1,200 textiles jobs annually (Benneworth & Hospers, 2007). By 1975, the industry completely lost its once dominant regional economic function, exposing the urgent shortage of in the region of innovative entrepreneurs creating new businesses.

The Drees Cabinet at a national level had introduced in 1951 the first regional economic policy in the Netherlands, although much of this became focused on reordering agricultural land holdings to allow increased food productivity and release land for economic development. From the early 1970s, the Dutch national economy faced a curious growth problem related to the two oil shocks, the so-called Dutch disease. Its substantial hydrocarbon reserves generated considerable state income: the first oil shock in 1970 led to an appreciation of its currency, which at the same time eroded its export competitiveness at the time global recession, was shrinking export markets. The second oil shock in 1973 unleashed a wave of economic problems for Dutch manufacturing, and in 1975, the OOM was created as a partnership between the national and provincial levels to deal specifically with Overijssel's structural economic problems, the decline of manufacturing. and create new jobs for those made unemployed by textiles' decline. The problem was an absence of clear foundations for regional growth: there were innovative businesses, including TenCate and the aerospace company Philips Signaal (today Thales), along with the regional university. The decline of textiles was indeed a problem for the university, which facing falling enrolments, faced parliamentary questions concerning its future existence. What was missing were individuals capable of creating new businesses, the next generation of 'barons' with the capital, ideas, and contacts to create new firms and generate significant numbers of jobs (cf. Willink, 2010).

EXPERIMENTS IN UNIVERSITY ENTREPRENEURSHIP AND INNOVATION NETWORKS

The first step of the analysis is creating a singular narrative of the changes which took place from the nadir of Twente's fortunes, marked by the oil crisis, the threatened closure of its university, and mass unemployment. Our argument is that a series of networks and concomitant cultures emerged since that point in which innovation and learning took place in Twente, but not at the scale at which it could be described as a learning region. Nevertheless, there are two main points in this narrative. The first is that there has been an evolution and growth in the scale of those networks, from their almost total absence in 1975 to their present state. Second, this evolution has not been straightforward, and part of that complexity can be explained by the dynamic tension between culture and innovation activities: understanding which activities have succeeded provides a useful insight into the underlying cultural dynamics. Third, it is still ongoing and has evolved into a very politically complex story: in order to simplify a complex situation, we have divided the change trajectory into three approximately equivalent time periods, in which there was a qualitative and visible shift in the nature of regional innovation networks.

Creation of a Community of Entrepreneurs (1979–1988)

Our argument is that the university was closely involved with the creation of a new class of engineers and entrepreneurs in Twente, and from the late 1970s onward, the university itself started to stress and manage this as part of trying secure its own survival. In 1978, Professor Harry van den Kroonenberg became Rector of THT, a professor of energy research who had had some success in encouraging his graduates to use their research projects to create new businesses. This gave him a model he later extended to the whole of the university, encouraging graduates to create new businesses using knowledge generated in university-based research projects, with this group of entrepreneurs justifying THT's existence by taking THT knowledge into society. Initially he created an Industrial Liaison Office (ILO) to help exploit the university's knowledge, allowing research groups to keep any income raised. At first, support for students' encouragement was purely moral, not throwing barriers up to students who wanted to commercialise their projects. But the concept developed in two complementary directions. First, one of the companies was a consultancy business who mapped THT's spin-off companies, and then mapped spin-outs from all Dutch universities for the Ministry of Economic Affairs (Van der Meer & Van Tilburg, 1984). This research helped to convince the Ministry to provide all universities with a 5-year grant to support spin-off activities by providing access to four resources, namely, finance, professorial expertise, corporate space, and business advice. THT used this funding to create a revolving fund which provided finance as a loan, and this fund formed the basis of the now-renowned TOP scheme (Temporary Entrepreneurs' Scheme, Tijdelijke Ondernemers Programma) which operated at first through the ILO.

The second development was the decision by OOM to create an incubator unit immediately to the south of the university campus, using a concept from the Control Data Corporation to create flexible rental units for high technology businesses, who together with ABN Amro and OOM were the main investors in the development. The building, called the Business Technology Centre (BTC), helped to support the sense that the graduates of the TOP programme were part of a community by offering a common space for their co-location. The risk for incubator units in declining industrial regions is that they become luxury office units rather than a hub of science-based businesses (cf. Massey et al., 1992), and the BTC put considerable effort into ensuring that its tenants were indeed science-based, or at least linked to the university, while ensuring a sufficiently high occupancy rate to cover the mortgage costs.

The effect was that BTC and the TOP scheme coincided in forming a community of high-technology entrepreneurs with close links to the university. The location provided the spin-offs with easy access to the university, and as the programme and BTC developed, provided a cohort of TOP alumni who were then capable of being enrolled as business mentors for the following sets of TOP participants. The boundaries of this community were blurred: the TOP programme was open to all companies, making even companies without direct spin-off links to the university part of a community in which both entrepreneurship and university engagement were everyday practices. Links between firms were at this stage largely informal, and there was a sense that more could be done to organise collective action among the community, which is what led to the formation of the Twente Technology Circle.

Pressures on the Community and the TKT (1988–1995)

The next phase in the evolution came with the retirement of Professor Van den Kroonenberg from his second Rectorial term (1988). THT had rebranded itself in 1987 as the "University of Twente: the entrepreneurial university" (UT) building on TOP's apparent success. Van den Kroonenberg's retirement threatened the commercial and innovative activities which UT was developing, and this stimulated both the ILO and the RDA to try to find new mechanisms by which UT could contribute to improving the region's high-technology base. 1988–1995 saw a number of experiments which attempted to expand the TOP scheme's success, and to sustain the sense that entrepreneurship was important to the university. There were both successes and failures in these projects, when regarded in terms of the longevity and growth of projects).

One big success from this period was the creation of the Twente Technology Circle (TKT, Technologie Kring Twente). A clear problem for co-operation between UT and its spin-offs was that there was a huge divergence of perspectives (and cultures) between a large organisation, its professors, and the micro-businesses emerging from the TOP scheme. The original idea behind TKT when it began in 1990 was to bring together groups of local entrepreneurs to form consortia who could approach large organisations in the region—whether the university, public organisations such as hospitals and housing corporations, and large firms—and sell collectively to these large firms. By doing this, the university liaison office hoped to create a set of well-configured industrial partners for their professors, which would in turn ensure a steady flow of new research projects with industrial partners, placements for their students, and ultimately, wider recognition for their regional engagement activities.

The original accent on TKT throughout this period, actively supported by both UT's liaison office and the OOM, was in enrolling a membership and introducing those members in the hope they would be able to co-operate. Interviewees reported that the fact that many companies had a common background through the TOP programme meant that they knew both the original TKT organisers as well as each other, and this provided a social lubrication that allowed the organisation to establish itself relatively successfully. Although the original ambitions to sell collectively were not fully realised, there was success in mobilising a membership of high-technology entrepreneurs. As with the BTC, not all these entrepreneurs were TOP firms, nor were all located in the BTC, but participated in a shared community in which contact and interaction with the university were seen as normal activities, and the university was supportive or at least tolerant of these activities. In this period, the municipality and university co-operated on the development of a science park in the land adjacent to the BTC, the rationale being to create a location for longer-lived TOP firms to grow into, and to create a new visible high-technology sector in the region. This indicates the extent to which the TOP community with its links to the university was building a degree of credibility in the policy community.

At the same time, interviewees reported many projects which were not successful, or at least did not lead to continuation activities (TOP) or growth (TKT) suggestive of a bigger cultural change. The most notable of these was the creation of an enterprise office as a trading fund to commercially exploit university assets, including developing a campus hotel. The trading arm closed suddenly after its commercial approach diverged from university accounting systems. Other projects were created, ran their course, and disappeared without leaving a visible trace: UT became involved in a number of entrepreneurship promotion schemes in a variety of places: these all operated and created a set of outputs, but unlike the TOP scheme, failed to establish themselves as part of the landscape of business support. There were also problems in using European Regional Development Funds (from 1989) to support regional innovation: a "Regional Technology Plan Twente" was developed by consultants, and the Twente Innovation Stimulation Programme was created. The former remained a paper strategy never to be implemented, and the latter was more an instrument for specific firm support than creating a wider community which became the basis for further interventions.

The Bursting of the High-Technology Bubble (1995–2004)

The situation after 1995 became much messier because of the increasing emphasis placed by the European Commission on regional innovation policy, and the increasing availability of funds to carry out 'innovation' projects. In the late 1990s, a number of business support agencies in Twente became entangled in the 'dot.com' bubble, an irrational belief that Internet businesses would thrive even despite the absence of the usual fundamentals such as an effective business plan. This had a disruptive effect on the region, as there was a contamination effect, with good projects, including the TOP, becoming embroiled in misconceived actions which undermined their validity and support. So in 2000, for example, there were national ICT entrepreneurship programmes called Dreamstart and Twinning, which ultimately disappeared without trace after the majority of their investments failed as businesses. But during this period, Twente's regional innovation culture developed in terms of an expanding community of innovators with links to the university. The first is that TKT managed to successfully mobilise a series of collaborative innovation activities between member firms, which saw the regional innovation network evolve in shape from being a hub-and-spoke around the university, to one with more nucleii, including TKT. Klein Woolthuis (1999) mapped one of these, the Twente Initiative for Medical Product development (TIMP), which began as a subsidy funded project, and during which time a number of its member firms grew considerably in size.

There were also other collaborations facilitated by the TKT identified by interviewees. There was a rash of 'Valley' organisations created around the turn of the century. These were inspired by the Silicon Valley brand, but the reality was a great deal more variable. The Mechatronica Valley Foundation was founded in 2004 by a number of TKT firms and helped to sponsor a Professorship within the University as well as organising an annual regional university/ industrial mechatronics conference. The Virtual Reality Valley evolved into the Virtual Reality Laboratory based on the university campus and formed part of the Technology Exchange Cell partnership between the University and Thales. The Membrane Valley Foundation was an attempt to mobilise firms including spin-offs and the university's membrane research activities into a collective organisation: this disappeared without trace because of a lack of value added in networking activities, although firms and the research group continued their successful existence.

There were also further developments within the university, including the emergence of the MESA+ incubator facility. This was created as a shared facility between a number of nanotechnology and materials science research groups and external businesses, allowing university research groups access to expensive kit and firms to access university knowledge. This environment was a good location for student dissertation research, and as a result, MESA+ came into consideration for the TOP programme, and in this period, we see around 20 new companies created within the MESA+ laboratories, the majority drawing on the TOP programme, and creating a new wave of high-technology businesses. Interviews with these companies suggested that MESA+ was an important part of their community identity, as well as the TOP programme. This MESA+ identity could be rationalised as emerging from their close links to other MESA+ companies which emerged socially trying to solve difficult nanotechnology problems drawing on the tacit knowledge of the university and corporate researchers active in the field. There were also a number of failures which put the successes such as TIMP and MESA+ into contrast. One of these was the Twente Knowledge Initiative, which sought to offer subsidies to groups of firms to collectively develop products: in the generally buoyant economy, firms were doing well, but after the economic tightening of 2001–2002, it became clear that very little sustainable co-operation in innovation had been created. The Province developed a regional innovation strategy (covering the whole of the Province and not just Twente), with a number of flagship initiatives, but telling about this was that although they were delivered, little trace of those activities remained even five years after the projects were completed, nor were they extensively talked about by interviewees as significant in the way that other things, notably TOP, were. A more positive effect of this was that there were clear signs that regional politicians were becoming more enthusiastic about the capacity of spin-offs to create new market sectors, something which became central in the following phase, which saw the development of the Kennispark (Knowledge Park project) which is as much a political flagship project as it is about supporting the development of networks of innovators.

SOCIAL CAPITAL AND LOOSELY COUPLED INNOVATION NETWORKS

First Cut Analysis of the Synthetic Narrative

From the synthetic narrative, there appear to be a number of key issues which emerged in distinguishing between activities which were successful, and part of a wider regional change and those which were unsuccessful and not part of this change process. The first was that despite this being a regional story, it was the innovators themselves who were critical to the process of change, and the social dimension to the individual story is important. Because of the huge emphasis that is placed on innovation policy, it can be made to seem that policy is what creates innovative environments. The key actors in this story are the entrepreneurs and innovators themselves. These were not always from the region, but were attracted to the region by the university or their employment. The TOP scheme provided an opportunity for these people to stay temporarily in the region. In the course of this personal innovation and entrepreneurship journey, they had the opportunity to participate in a community of innovators. Policy encouraged a continual stream of new members for this community, and continually offering resources—loan, contacts, bench space, business advice—to help these innovators create new high-technology businesses. Everything that emerged in the case of Twente built on these innovators, each engaged in his or her own competitive corporate struggles. The second element was one over-riding similarity between the innovators around the university, providing a shared experience, and driver of mutual connectivity, and that was very high external dependencies. These innovators were almost all attempting to innovate in the context of a micro-business with no established cash flow, tangible assets, or existing product. This made the innovators dependent on resource rich partners: the university became a source of knowledge and space resources, while other companies became a source of knowledge, expertise, and business mentoring. The university, and later TKT, provided the spaces where innovators could meet to access these resources: the networks were not obligatory for those high-technology firms, but those successful networks did help those innovators to access innovation resources. Participating in the networks was both useful to innovators in Twente, and it also became a behavioural routine in the network.

The third element of the story was its constructive nature: successful policy interventions were the ones that took an existing (potentially latent) strength embedded within a network of innovative actors, and qualitatively extended it, in terms of numbers of participants, strength of interactions, number of interactions, or the external profile and reputation of the scheme. The TOP scheme exemplifies this constructive process: it involved the formalisation of an informal approach already tolerated around the university, but extended its resource provision (providing participants with a loan to cover salary costs) and led to an increase in numbers over time (Van der Meer & Van Tilburg, 1984; van der Sijde et al., 2002a; 2002b). Likewise TKT took the mentoring that took place between cohorts of the TOP programme to create more regular connections between business with the result being a number of concrete innovation networks. Mechatronica Valley and TIMP created networks of TOP companies who knew one another already and created collaborative innovative activities between them.

What can be seen here in terms of the development of an innovative or learning culture is a development in the nature of innovation networks by which innovators access resources from the immediate and personal to the more imagined and epistemic, evidenced in a formalisation and routinisation of the activities. At the same time, the direct effect on innovation remains through concrete and direct connections between innovators. So the learning culture has evolved in the sense of there being a step change in the qualitative magnitude of co-ordination and connectivity between actors. The change has been one of direction of travel rather than an absolute shift; in the quarter-century under consideration, Twente has evolved from being a place from where innovation is rather hard to achieve to one where it is much easier to make useful connections to access scarce resource. It is important here not to make a totalising claim that all innovation is easier in Twente, nor that the region has become more innovative. Nevertheless, the development of these networks and their success in creating new businesses have seen evolving innovation networks interacting with the wider regional culture. Twente's regional culture may thus be argued to facilitate learning in regional networks but that does not mean that all regional learning in Twente benefits from the region's culture in an equal way. That very much depends on the regional networks and how they engage with the wider regional culture, thereby underscoring that the appropriate level of analysis to look at for an explanation of regional culture may benefit learning and innovation is thus regional networks, not the region as such.

The Dynamic of Interaction between the Different Elements

This analysis provides a means to answer the three research questions initially raised, and hence our overall research question. The first question asked was what were the innovation resources which innovators accessed using social networks. The answer to this question was that there was not a specific class of innovation resources associated with network activity. The issue in Twente for innovating businesses was that all resources were scarce, and networks provided a means to access resources for innovation not otherwise readily available. The experience of innovating in Twente was of negotiating scarcity and this became associated with a rather high degree of risk aversion, which some commentators tied to a more general cultural desire for lifestyle businesses, but on the other hand reflects a reality that successful regional innovators had often achieved a very deal with limited resources. This created this very cautious regional innovation 'style' which intuitively echoed the longer-standing regional culture of austerity without being directly caused by it. Nevertheless the fact that there was an innovation culture can be regarded as a shift, but it is clear in talking about a regional innovation culture there is only reference made to a group of people, and not as if the region made innovators unambitious.

This brings us to the second question, which is how do the norms and dynamics of those innovation networks relate to the wider local cultures. What is visible in the Twente case study is the emergence of plural agency, what we highlight as the multiple nodes of activity. What started as a set of activities closely anchored around the university became driven by a cohort of local entrepreneurs. The spinning off of the TKT from the university and the creation of TIMP and Mechatronica Valley are examples of strong regional innovation leadership emerging in the private SME sector. It would clearly be ridiculous to claim this as a complete cultural shift, but this new entrepreneurial leadership of innovation activities is associated with a new sub-culture in the region, that of high-technology innovator. This sub-culture is still vulnerable and difficult to organise structurally, but the emergence of this sub-culture again points to a change in the composition of the regional culture and its overall innovative capacity.

This leads neatly to the answer to the third question, which asked how the institutionalisation and assimilation of these loose innovation networks into a wider regional culture take place. The cultural change can be regarded as the emergence of a new group—the sub-culture of innovative entrepreneurs—who are both materially and symbolically powerful. They are materially powerful because they are rich, and they are symbolically powerful because they represent a necessary transformation which local and regional politicians are keen—for their own divergent reasons—to be able to claim as their own. The breadth and range of publicly funded innovation projects seeking to support innovators in Twente indicate the depth of this symbolic power and taken together these highlight that the change in culture that has emerged because of the emergence of this novel powerful group. The consequence of this is that we are not talking about a regional demotic culture based around an imagined community, but rather a regional elite culture of innovation, that cannot be divorced from the exercise of social power.

TOWARD A BETTER SPECIFIED TERRITORIAL INNOVATION NETWORK RESEARCH AGENDA

In this chapter we have been concerned to address the question of how do the social habits, norms, and routines of local innovation networks shape the wider culture and behaviour of their particular local territory, and shape the local innovation environment as well as economic development outcomes. A key issue in this chapter has been the scalar dialectic in these analyses. The literature is clearly divided by the conceptual desirability of avoiding the reification of regional culture as a meso-scale actor but the practical necessity of talking about regions in ways that may suggest reification. The question raised in this chapter is to what extent does considering social network dynamics allow this tension to be bridged, and how can social network analytic frameworks for innovation incorporate micro-features but still have something sensible to say about territory and regions. The key issue here is the dependency of innovators on easily-accessible resources to successfully innovate in a less successful region (cf. Sheamur, 2011). Some innovators have been able to access resources outside the region, and indeed one of the contributions made by the university is facilitating access to external knowledge resources, but the point here is one of contingency that is the need to repeatedly access resources that are structurally absence.

This brings us back to the importance here of dynamics and change, rather than thinking about territory in terms of structure and boundaries. What is important is how a single innovator can access resources at the opportune moment: the innovator has a resource dependency that is in part shaped by territorial context and positionality in social networks at that time. This suggests that rather than trying to make statements about structural change at the meso-level of the region, the link between region and innovator is an emergent property. Here we are struck by the fact that when so conceived, regional culture is something that primarily applies to the individual innovators—following Granovetter's prescription, collective groups shape individuals without binding them. Yet, at the same time, regional innovation studies and policy are focused primarily on the collective innovation network or formal institution, thereby creating an elision between regions and networks A useful first conceptual step would be to separate out two kinds of networks, individuals in their wider social networks that create latent potential, and formal collective innovation relationships that create direct economic benefits.

In making this distinction, one can look at the differential effects of territory on the dynamics of those networks, and bring the 'region' back to regional innovation without the problem of reification. Here we see potentially value in drawing on Burt's distinction in network analysis (1980) between individual ego networks (individual as a hub with spokes to other connections) and collective social networks (the aggregate of ego network connections in a single domain). The key distinction is that ego networks are multiplex while social networks are simplex: individuals have relationships in multiple contexts while social networks are purposive and relate to a reason to interact (Knoke & Yang, 2008). Our contention here is that it is individuals which are sticky and connected to territory, and therefore it makes sense to focus analysis on individual innovators' ego networks, how individuals relate to territory, and how that embeds the innovation in particular places. This allows the question to be asked of which are the important social and cultural relationships which tie people to a particular territory in terms of innovation processes, and how can we conceptualise how these individual cultural dynamics and connections at a regional level, necessary to derive sensible practical lessons for better managing regional innovation activities.

The individual innovator becomes understood as a hinge or bridge between two different network worlds, each of which have quite different relationships with the territory, and which together influence the developments of those networks and ultimately economic development in those territories. There also seem to be echoes in some of the emerging work in economic geography around mobility of highly skilled people and their non-functional cultural choices (e.g., Florida, 2008). In this sense, territoriality in innovation networks becomes more about the relationships of the individuals to places, and their cultural and social affiliations in these places, rather than a function of the particular innovation networks and formal institutions which emerge in particular places, and for which a relationship with wider regional culture is claimed. Our plea is therefore for more research to focus on this issue of how regional 'culture' becomes operationalized and influential in individuals' ego networks, influencing in turn their innovation performance, and allowing a much more rigorous approach to understanding the effect of territory on innovation in networks.

ACKNOWLEDGMENTS

An earlier version of this chapter was presented at the Roundtable session "The Social Dynamics of Innovation Networks" at the Regional Studies Association International Conference, Newcastle upon Tyne, United Kingdom, April 18–20, 2011, and the authors are grateful to all Roundtable participants for their insightful comments. The empirical part of this paper reports findings from a series of research projects into the development of an entrepreneurial culture in the region of Twente funded from a variety of sources, including the U.K. Economic and Society Research Council project "Bringing Cambridge to Consett," the OECD IMHE programme Universities and Regional Development, an Institute for Innovation and Governance Studies Visiting Fellowship at the University of Twente, a Research Councils U.K. Fellowship on the Territorial Dimension of Innovation and the Center for Higher Education Policy Studies. All errors and omissions remain the responsibility of the authors.

NOTES

1. We will not concern ourselves with one set of critiques, that of irrelevance, which has come from approaches which regard territorial unevenness as a form of spatial disequilibrium within self-equilibrating systems, and hence necessarily temporary and a distraction.

REFERENCES

Albrechts, L., & Swyngedouw, E. (1989) The challenges for regional policy under a flexible accumulation regime. In L. Albrechts, F. Moulaert, P. Roberts, & E.

Swyngedouw (eds.), *Regional policy at the crossroads: European perspectives*. London: Jessica Kingsley.

Amin, A., & Palan, R. (2001) Towards a non-rationalist international political economy. *Review of Iinternational Political Economy*, 8, 559–577.

Amin, A., & P. Cohendet. 2004. *Architectures of knowledge: Firms, capabilities, and communities*. Oxford: Oxford University Press.

Asheim, B.T. (1996) Industrial districts as "learning regions": A condition for prosperity. *European Planning Studies* 4 (4), 379–400.

Benner, C. (2003) Learning communities in a learning region: The soft infrastructure of cross firm learning networks in Silicon Valley. *Environment & Planning, A* 35 (10), 1809–1830.

Benneworth, P.S., & Henry, N. (2004) Where is the value added in the cluster approach? Hermeneutic theorizing, economic geography and clusters as a multi perspectival approach. *Urban Studies*, May 2004 vol. 41 no. 5–6 1011–1023.

Benneworth, P.S., & Hospers, G.J. (2007) Urban competitiveness in the knowledge economy: universities as new planning animateurs. *Progress in Planning*, 23 (1), 3–102.

Bijleveld, P., & Geerdink, C. (2010) Twente Index 2010 Enschede: Stichting Twente Index. http://www.twente-index.nl/lotusroot2/3000/CMStwenteindex2010.nsf/(Design)/TwenteIndex2010.pdf/$FILE/TwenteIndex2010.pdf, accesed July 23).

Boschma, R.A. (2005) Proximity and innovation. A critical assessment. *Regional Studies*, 39 (1), 61–74.

Buck Consultants. (2010) *Fysieke investeringsopgaven voor campussen van nationaal belang*. The Hague: Ministry of Economic Affairs (EZ).

Buesa, M., Heijs, J., & Baumert, T. (2010) The determinants of regional innovation in Europe: A combined factorial and regression knowledge production function approach., *Research Policy*, 39 (6), 722–735.

Burt, R.S. (1980) Models of network structure. *Annual Review of Sociology*, 6, 79–141.

Butzin, B. (2000) Netzwerke, Kreative Milieus und Lernende Region: Perspektiven für die regionale Entwicklungsplanung. *Zeitschrift für Wirtschaftsgeographie*, 44 (3/4), 149–166.

Christopherson, S., & Clark, J. (2007) Power in firm networks: What it means for regional innovation systems. *Regional Studies*. 41 (9). 1223–1236.

Christopherson, S., & Clark, J. (2010) Limits to 'The Learning Region': What university-centered economic development can (and cannot) do to create knowledge-based regional economies. *Local Economy*, 25, 120–130.

Clark, B. (1998) Creating entrepreneurial universities: Organizational pathways of transformation. Oxford: Pergamon/IAU Press.

Coenen, L. (2006) "Faraway, so close" the changing geography of regional innovation. *CIRCLE*, Lund University, Lund, Sweden.

Cooke, P. (2005) Regionally asymmetric knowledge capabilities and open innovation : Exploring 'Globalisation 2'—a new model of industry organisation. *Research Policy*, 34, 1128–1149.

De Vries, L. (2005) Een wereld te winnen: Koninklijk Ten Cate en recente ontwikkelinge in de wereldwijde textiel industrie. *Textielhistorische Bijdragen*, 45, 52–58.

Florida, R. (2008) *Who's your city*. New York: Basic Books.

Gertler, M.S., & Wolfe, D.A. (2006) Spaces of knowledge flows. Clusters in a global context. In B. Asheim, P.N. Cooke, & R. Martin (eds.), *Clusters and regional development. Critical reflections and explorations*. London: Routledge.

Gordon, I.R., & McCann, P. (2000) Industrial clusters: complexes, agglomeration and/or social networks? *Urban Studies* 37 (3), 513–532.

Granovetter M. (1985) Economic action and social structure: The problem of embeddedness. *American Journal of Sociology*, 91 (3), 481–510.

Haas, Peter M. (1992) Epistemic communities and international policy coordination. *International Organization*, 46 (1), Winter (MIT Press), 1–35.

Harrison, B. (1992) Industrial districts: Old wine in new bottles? *Regional Studies*, 26 (5), 469–483.

Hassink, R. (1993) Regional innovation policies compared. *Urban Studies*, 30 (6), 1009–1024.

Hassink, R. (2007) Learning regions: A constructive critique. In R. Rutten & F. Boekema (eds.) *The learning region; Foundations, state of the art, future*. London: Edward Elgar, 252–272.

Hassink, R. (2010) Regional resilience: A promising concept to explain differences in regional economic adaptability? *Cambridge Journal of Regions, Economy and Society* 3 (1), 45–58.

Hassink, R, &Lagendijk, A. (2001) The dilemma of inter-regional institutional learning. *Environment and Planning C: Government and Policy*, 19, 65–84.

Hudson, R. (2003) Fuzzy concepts and sloppy thinking: Reflections on recent developments in critical regional studies. *Regional Studies*, 37, 741–746.

Irawati, D., & Rutten, R.P.J.H. (2011) The Java automotive industry: Between keiretsu and learning region. *Journal for Global Business Advancements*, 4(3), 208–223.

Karlsson, C. (2008) Introduction. In C. Karlsson (ed.), *Handbook of research on innovation and clusters: Cases and policies*. Cheltenham: Edward Elgar.

Klaassen, L.H. (1968) *De functie van Twente in de Nederlandse economie: met beschouwingen betreffende Oost Gelderland en Westmünsterland*. Rotterdam: Nederlands Economische Instituut.

Klein Woolthuis, R. (1999) Sleeping with the enemy: Trust dependence and contract in interorganisational relationships. Enschede, the Netherlands: University of Twente Press.

Knoke, D., & Yang, S. (2008) *Social network analysis*, 2nd ed. Thousand Oaks, CA: Sage.

Koschatzky, K. (2009) *The uncertainty in regional innovation policy: Some rationales and tools for learning in policy making*. Working Papers Firms and Region No. R6/2009. Karlsruhe: Frauenhofer ISI.

Lagendijk A. (2003) Towards conceptual quality in regional studies: The need for subtle critique—A response to Markusen. *Regional Studies*, 37 (6/7), 719–727.

Lagendijk, A. (2006) Learning from conceptual flow in regional studies: Framing present debates, unbracketing past debates. *Regional Studies*, 40 (4), 385–399.

Lagendijk, A. (2007) The accident of the region: A strategic relational perspective on the construction of the region's significance. *Regional Studies*, 41 (9), 1193–1208.

Lagendijk, A., & Cornford, J. (2000) Regional institutions and knowledge—Tracking new forms of regional development policy. *Geoforum*, 31, 209–218.

Lagendijk, A., & Oïnas, P. (2005) Proximity, external relationships and local economic development. In A. Lagendijk & P. Oïnas (eds.), *Proximity, distance and diversity, Issues on economic interaction and local development*. London: Ashgate.

Lazzeretti, L., & Tavoletti, E. (2005) Higher education excellence and local economic development: The case of the Entrepreneurial University of Twente. *European Planning Studies*, 13 (3), 475–492.

Lorenzen, A. (2008), Knowledge networks in local and global space. *Entrepreneurship and Regional Development*, 20 (6), 533–545,

Lovering, J. (1999) Theory led by policy: The inadequacies of the 'new regionalism' (illustrated from the case of Wales). *International Journal of Urban and Regional Research*, 23 (2), 379–395.

Malecki, E. (2010) Global knowledge and creativity: New challenges for firms and regions. *Regional Studies,* 44 (8), 1033–1052.

Markusen, A.R. (1999) Fuzzy concepts, scanty evidence, policy distance: The case for rigour and policy relevance in critical regional studies. *Regional Studies,* 33 (9), 869–884.

Martin, R., & Sunley, P. (2003) Deconstructing clusters: Chaotic concept or policy panacea? *Journal of Economic Geography,* 3 (1), 5–35.

Massey, D., Quintas, P., & Wield, D. (1992) *Hi-technology fantasies.* London: Routledge

Van der Meer, J. D., & Van Tilburg, J.J. (1984) Spin-offs van de Nederlandse kenniscentra, Economisch-Statisti- sche Berichten, 69 (3485), 1170–1173.

Morgan, K. (1997) The learning region: Institutions, innovation and regional renewal. *Regional Studies,* 31 (5), 491–403.

Morgan, K. (2004) The exaggerated death of geography: Learning, proximity and territorial innovation systems. *Journal of Economic Geography,* 4 (1), 3–21.

Moulaert, F., & Nussbaum, J. (2005) Beyond the learning region: The dialectics of innovation and cultural in territorial development. In R.A. Boschma & R.C. Kloosterman (eds.), *Learning from clusters: A critical assessment from an economic-geographical perspective.* Dordrecht: Springer.

Moulaert, F., & Sekia, F. (2003) Territorial innovation models: A critical survey. *Regional Studies,* 37 (3), 289–302.

Nuur, C. Gustavsson, L., & Laestadius, S. (2009) Promoting regional innovation systems in a global context. *Industry & Innovation*, 16 (1), 123–139.

Oosterlynck, S. (2007) The political economy of regionalism in Flanders: Imagining and institutionalising the Flemish regional economy. Unpublished Ph.D. thesis, Lancaster: University of Lancaster.

Pedler, M., Burgoyne, J., & Boydell, T. (1991) *Towards the learning company: Concepts and practices.* London: McGraw-Hill.

Piore, M.J., & Sabel, C.F. (1984) *The second industrial divide.* New York: Basic Books.

Romer, P.M. (1986) Increasing returns and long-term growth. *Journal of Political Economy,* 94 (5), 1002–1037.

Rutten, R., & Gelissen, J. (2010) Social values and economic development. *European Planning Studies*, 18 (6), 921–940.

Sheamur, R. (2011) Innovation, regions and proximity: From neo-regionalism to spatial analysis, *Regional Studies*, 45 (9), 1225–1243.

Sijde, P. van der, Karnebeck, S., & Benthem, J. v.,(2002a) The impact of a university spin off programme: The case of HTSFs established through TOP. In R. Oakey, W. During, & S. Kauser, (eds.), *New technology based firms in the millennium,* volume 2. London: Pergamon.

Sijde, P.C. van der, Vogelaar, G., Hoogeveen, A., Ligtenberg, H., & Velzen, M. van (2002b) Attracting high-tech companies: The case of the University of Twente and its region. *Industry and Higher Education*, 16, 97–104.

Storper, M. (1993) Regional "worlds" of production: Learning and innovation in the technology districts of France, Italy and the USA. *Regional Studies,* 27 (5), 433–455.

Storper, M. (1995) The resurgence of regional economies ten years later: The region as a nexus of untraded interdependencies., *European Urban & Regional Studies* 2 (3), 191–221.

Trigilia, C. (1991) The paradox of the region: Economic regulation and the representation of interests. *Economy and Society,* 20 (3), 306–327.

Uyarra, E. (2009) What is evolutionary about 'regional systems of innovation?' Implications for regional policy. *Journal of Evolutionary Economics*, DOI 10.1007/s00191-009-0135-y.

Van der Meer, J D., & Van Tilburg, J. J. (1984) Spin-offs van de Nederlandse kenniscentra, 69 (3485), pp.1170–1173.

Wenger, E. (1998) Communities of practice: Learning, meaning and Identity. Cambridge, New York, Melbourne: Cambridge University Press.

Westlund, H., & Adam, F. (2010) Social capital and economic performance: A meta analysis of 65 studies. *European Planning Studies*, 18 (6), 893–920.

Willink, B. (2010) *De textielbaronnen: Twents-Gelders familisme en de eerste grootindustrie van Nederland, 1800–1980*. Zutphen (NL): Walburg Pers.

Discussion and Conclusion

Instead of a Conclusion
Society, Culture, and Innovation—Themes for Future Studies

Frane Adam and Hans Westlund

The basic argument for compiling a book about the socio-cultural contexts of innovation is that innovation does not depend solely on 'pure' technology and the economy, i.e., the technological know-how and resources (R&D expenditure and financial and human capital) to exploit it. Both technology and the economy exist in societies whose culture influences preferences and behaviour. Culture sets (often very necessary) limits on the use and development of technology and the economy—but it can also create favourable conditions for new combinations of knowledge (and other resources) that result in what are called innovations.

Such a perspective does not assert that culture is the sole driving force of the success of a place or a region. Instead, from Marshall's air and Weber's Protestant ethic through to Florida's creativity there is an (often implicit) recognition of factors like technology/R&D, human capital, financial resources, production factors' relative costs, infrastructure, policies, etc., which all set the stage for innovation, development, and growth. But then, culture and social capital also matter as they affect the relations and interactions between those agents who have these 'traditional' resources available. This important, but often neglected aspect of innovation lies at the core of the contributions in this book. Westlund's and Li's chapter presents a typology of various dimensions of collaboration in innovation systems and the role of social capital in these collaborations.

If we broaden this approach outside the defined, policy-supported innovation systems to society and the economy in general, it is obvious that we need more knowledge on how cultural and other 'soft' (intangible) factors like social capital influence actors' behaviour in interaction and collaboration with others; how preferences, norms and values affect their willingness and ability to learn and combine and transform knowledge into new knowledge. Are some preferences, norms, and values more important than others when it comes to facilitating (or preventing) collaboration and creative learning? We need new hypotheses—and research—on how culture and social capital influence these processes.

So far our knowledge is quite limited and rudimentary. Empirical research on these issues is still in the pilot phase; the findings are relatively inconclusive. In this volume several contributions seek to synthesise the new cognitions as well as conceptual and methodological shortcomings in certain fields/topics of research. Proceeding from these insights (referring to the contributions by Adam, Chapter 5; Rutten & Irawati, Chapter 7; Benneworth & Rutten, Chapter 9; and others), it can be argued that culture and social capital do matter for innovation performance. This means that it makes sense to continue investigating this interaction. However, there are many problems and open dilemmas regarding the issues of generalisation, metrics (indicators), sampling as well as reliability when using data from just one dataset. We know that in many papers author(s) typically conclude with the statement "more research is needed." Yes, this is true, but the key question is: what kind of research? Should we continue with the old research design which has substantial shortcomings or should the previous research design be replaced/modified by a new research strategy? We favour the latter option (Adam's chapter provides detailed instructions on which changes are needed at the levels of methodology and organisation of the research process). We want to highlight the themes and accents for future research; themes that both are 'in the air' of regional economic theory and economic geography as well as (economic) sociology and are also suggested by authors in this volume.

On a more concrete level, two findings relating to the social capital–innovation interaction should be pointed out. The first concerns the distinction between bonding and bridging capital. The literature shows that many authors consider bonding social capital (circles or networks with relatively strong ties) as factor hampering innovativeness and bridging capital as a factor promoting innovativeness. Our analyses support the argument that both types of social capital are important and can be productive. In the initiation phase of an innovation process (invention) exposure to different actors, ideas, and information may have a positive impact. However, in some later (implementational) phases, collegial support and close cooperation are essential. Here it is important to mention the distinction between bonding and bridging capital as the weak–strong tie dichotomy seems to be too crude, too schematic. We have to develop a more fluid, multi-dimensional typology.

Greater attention should also be paid to the fact we encounter different spatial layers of social capital (as well as cultural values). An (analytical) distinction must be made between the societal (macro) level of social capital on the national or regional level. This spatial layer can also be labelled contextual or generalised social capital. On the other side, we can have more specific and specialised social capital on the level of a firm, network, or team. Proceeding from the theoretical standpoint it can be assumed that a positive inter-correlation exists between these spatial levels. However, this is a matter of empirical verification and in some cases no such correlation can be found.

The second theme we would like to address affects the notion of networks. We know how often this notion appears in different semantic contexts. Researchers who are unfamiliar with network analysis use this term in a somewhat unspecified manner. In addition, not every kind of networking is beneficial for innovativeness and knowledge-sharing. Sometimes one can gain the impression that is the quantity of ties (links) that matters. For some occasions and frameworks it is better to use other terms like epistemic community or community of practice (see Roberts's Chapter 4 in this volume). Some (large) network-based research could also lead to the conviction that it is only structures that matter and that the actors are statists or passive recipients of information circulating across the network.

Both Rutten and Irawati and Benneworth and Rutten call attention to the reification and localism in regional networks as they have been applied in the economic geography framework. Chapter 8 by Ivančič et al. shows how we can gain insights from research on individual inventors and highlights the potential of more of such studies of other countries with different socio-cultural contexts. The theoretical importance of such an individual turn can be expressed like this: In social capital research, a graphical device frequently used is a network diagram that has circles or dots for nodes and lines for interactions or ties. It is very, very familiar, but it is fair to say that the dots are often black boxes (or 'black dots'). Research has mainly concentrated on the arrows and overall structure of the network, and has neglected the dots—the individual actors and their norms, values and preferences, their variety, historical experience, motivations to take part in the networks, etc.

The third accent seems to be perhaps the most important regarding future research on the socio-cultural context of innovativeness. It relates to the research methodology, metrics, and coordination of research projects dealing with these topics on the European level. The overviews and meta-analyses presented in this volume indicate the relatively fragmented, uncoordinated, and introverted approach taken by many authors and researchers. Without going into details about the reasons for such an approach, we use the opportunity to propose some alternative solutions. What we have in mind is the establishment of a European interdisciplinary research forum to study the socio-cultural dimensions of innovation performance. The main task of this forum should be to encourage discourse and reflection among researchers interested in these studies. It is believed that the exchange of ideas and search for the best methodological solutions will result in productive co-opetition (a combination of competition and cooperation). It is understandable that this cannot be imposed from above since this can only be a bottom-up initiative and we hope that this volume is regarded as one of the first steps in this direction.

Contributors

Alja Adam is a Researcher at the Institute for Developmental and Strategic Analyses (IRSA) and Assistant Professor at the School of Advanced Social Studies, Nova Gorica, Slovenia.

Frane Adam is Head of the Research Centre at the Institute for Developmental and Strategic Analyses (IRSA), Ljubljana, Slovenia; Professor at the Faculty of Social Sciences, University of Ljubljana, Slovenia; and Professor at the School of Advanced Social Studies, Nova Gorica, Slovenia.

Marian Adolf is Assistant Professor at the Karl Mannheim Chair for Cultural Studies at Zeppelin University, Friedrichshafen, Germany.

Paul Benneworth, FeRSA (Fellow of Regional Studies Association), is a Senior Researcher at the Centre for Higher Education Policy Studies at the University of Twente.

Pedro Ferreira is a Professor at IPAM—The Marketing School and Lusiada University, Lisbon, Portugal.

Ana Hafner is a doctoral candidate at the School of Advanced Social Studies, Nova Gorica, and Deputy Director of Association of Slovene Inventors ASI—Active Slovene Inventors.

Dessy Irawati is a Teaching Associate at Newcastle University Business School, United Kingdom.

Angela Ivančič is a Researcher at the Instututite for Developmental and Strategic Analysis (IRSA) and a contracted Assistant Professor at the Faculty for Social Sciences, University of Ljubljana, and the Faculty of Applied Business and Social Studies DOBA in Maribor, Slovenia.

Yuheng Li is Research Associate at the Institute of Geographic Sciences and Natural Resources Research, Chinese Academy of Sciences, Beijing, China.

Contributors

Jason L. Mast is a Research Fellow at the Karl Mannheim Chair for Cultural Studies at Zeppelin University, Friedrichshafen, Germany.

Isabel Neira is a Lecturer in Econometrics at the Faculty of Economics, University of Santiago de Compostela, Spain.

Darka Podmenik is Researcher at the Institute for Developmental and Strategic Analyses (IRSA), Slovenia.

Toni Pustovrh is a doctoral candidate at the Center for the Social Studies of Science, Faculty of Social Sciences, University of Ljubljana, Slovenia.

Joanne Roberts is a Professor in the Strategic Management and International Business Subject Group, Newcastle Business School, United Kingdom.

Roel Rutten is Assistant Professor in the Department of Organizational Studies at Tilburg University, The Netherlands.

Nico Stehr is Professor at the Karl Mannheim Chair for Cultural Studies at Zeppelin University, Friedrichshafen, Germany.

Elvira Vieira is based at the Portuguese Institute of Marketing Management.

Hans Westlund is Professor at KTH (Royal Institute of Technology), Stockholm, Sweden; JIBS (Jönköping International Business School), Jönköping, Sweden; and IRSA (Institute for Developmental and Strategic Analyses), Ljubljana, Slovenia.

Index

A
Adam, A. 161, 215
Adam, F. 2, 12, 15, 17, 19, 112, 115, 121, 124, 125, 144, 158, 162, 163, 171, 173, 181, 184, 208, 212, 215
Adler, P. S.83, 96
Adolf, M.2, 14, 25, 215
agency 32, 125, 181, 190, 191, 202
Albrechts, L. 188, 204
Aldhous, P. 46, 54
Amin, A.81, 82, 88, 91, 92, 95–97, 98, 100, 149, 156, 188, 189, 205
Ananda, R.40, 45, 54
Andersson, 21, 132, 140, 180, 181
Arefi, M. 45, 54
ASI—Association of Slovene Inventors 166, 169, 173, 176, 177, 215
Austin, J.L.30, 31, 38
automotive industry 16, 143, 145, 147, 148, 157, 188, 206

B
Bainbridge, W. S.41, 42, 47, 51, 54, 55, 56
Baker, W.142, 149, 150, 156, 173, 182
Bathelt, H., 88, 95, 97
Beck, U.45, 51, 54
Benner, C.80, 81, 89, 97, 185, 205
Best, M.142, 143, 144, 148, 156
Bijleveld, P. 192, 193, 205
biotechnology 35, 42, 43, 53, 55, 56, 66, 98, 184
Boschma, R.A. 88, 95, 97, 108, 123, 185, 187, 190, 205, 207
Bostrom, N.42, 50, 54
boundary processes 86–88
Bourdieu, P. 33, 38, 105, 163, 164, 182, 184

Brown, J. S. 20, 80, 84, 85, 88, 89, 96, 97, 169
Burger-Helmchen, T. 94, 97
Burt, R. S. 135, 140, 142, 144, 150, 183, 203, 205

C
Canton, J. 40, 54
case study 16–18, 89, 119, 135, 143, 145, 156, 161, 163, 165, 177, 184–186, 191–193
Castilla, E.J. 135, 140
Christopherson, S. 188, 189, 205
Church, A. H. 36, 38
civil
   society 1, 2, 6, 7, 8, 17, 37, 38, 39, 48, 49, 54, 105, 115, 133, 140, 164, 165, 170
   organisation(s) 14, 41, 52
Codes of ethics and conduct 46, 51
Coenen, C. 189, 205, 40, 47, 51, 54, 107, 123
Cognitive
   mobilisation 2, 3
   proximity 94, 95, 185, 190
Coleman, J.S. 80, 99, 103, 164,180, 182
collaboration 36, 49, 98, 107, 124, 126–141
Collins, H. M. 36, 38
community(-ies) of practice 15, 80, 81, 83–89, 92–100, 185, 208, 213
competencies 14, 25, 26, 27, 32–34, 181, 166
constellations of practice 86, 88, 98
controversial reception 104, 105
converging technologies 14, 40, 42, 54, 55
Cooke, P. 106, 107, 123, 157, 162, 164, 182, 188, 205

craft/ task-based 92
creativity 3–5, 7–9, 14, 21, 26, 28, 60, 64, 66, 79, 80, 83, 89, 92, 95, 97–100, 120, 123, 132, 151, 154, 155, 162, 163, 166, 168, 169, 174–176, 207, 184, 211
culture 10, 11, 12, 13, 17, 18, 19–21, 25
cultural
　dimensions 13, 15, 57, 59–61, 63–69, 71, 73–75 , 77, 78, 120, 125, 184, 213
　turn 10, 13

D
Dahlin, K. 161, 182
DeFillippi, R. 80, 89, 98
deliberation 41, 45, 46, 47, 48–53, 13
democracy 12, 19, 20, 27, 28, 39, 53, 55, 133, 140, 141, 184
Dewey, J. 47, 54
distributed communities 82, 83, 87, 88, 95
diversity 8, 10, 45, 79, 121, 151, 154, 162, 183, 206
Doner, R. 147, 156
Dosi, G. 38, 26
Drucker, P.F. 36–38
Durkheim, E. 36, 38, 144, 156

E
economic
　creativity 21, 64, 66, 79, 97, 98, 99, 100
　development 1, 2, 7, 8, 17, 20, 21, 27, 34, 39, 50, 51, 53, 60, 75, 89, 119, 123, 125, 128, 149, 150, 158, 186, 188, 193, 194, 202, 204–207
Ethical, Legal and Societal Implications (ELSI) 47, 50, 51, 183
empiricist 13, 113, 116, 122
entrepreneur 2, 4, 8, 9, 12, 13, 15, 19–21, 25, 28, 35, 43, 65, 75, 79, 89, 94, 114 , 117, 123, 124, 140, 145, 157, 162, 165, 166, 168, 170–172, 180–182, 193–198, 200, 202, 204–206
epistemic turn 110, 115
epistemic/creative 91, 92
Etzioni, A. 81, 82, 98
Etzkowitz, H. 2, 19
European Commission 28, 38, 54, 55, 198

Evans, R. 36, 38
Expenditure 71–74, 78, 118, 120, 211

F
Faulkner, W. 35, 38
Fereirra, P. 12, 14, 57, 215
firm(s) 6, 8, 97, 119, 120, 130, 134, 144, 156, 205, 206
Fleming, L. 108, 123, 101, 109, 182, 124
Florida, R. 7, 8, 19, 121, 123, 133, 140, 142, 150, 151, 156, 204, 205, 211
Foreign Direct Investment (FDI) 146, 157
Foss, N. J. 80, 98
Freeman, C. 123, 140, 41,180, 183
Fukuyama, F. 53, 54

G
Garreau, J. 45, 54
Gehlen, A. 36, 38
Geroski, P. 26, 38
Gertler, M. S. 81, 88, 95, 98, 188, 205
Gibbons, M. 2, 20, 26, 27, 38, 129, 140, 141
Glazer, S. 42, 51, 54
globalization 46, 58, 96, 182, 187, 205
Godin, B. 26, 36, 38
Grabher, G. 140, 149, 156
Grupp, H. 126, 141

H
habit 17, 186, 202
habitual 28, 30–32, 34
habitus 93, 99
Hafner, A. 161, 173, 182, 215
Handy, C. 80, 99, 177
Gertler, M. S. 81, 88, 95, 98, 188, 205
Hassink, R. 156, 188, 189, 206
Hauser, C. 112, 113, 118, 124, 144, 149, 156, 163, 179, 181, 183
Hofstede 10–14, 19, 20, 57, 58, 60–6, 66–68, 71, 74, 77–79, 94, 99, 120
Honda 146, 147, 156
Howells, J. 81, 99
Hudson, R.143, 152, 156, 188, 206
Hughes, J. J. 42, 55
human
　capital 2, 3, 9, 18, 63, 105, 111, 117, 120
　resources 127, 11, 16, 98

I

IFIA—International Federation of Inventors' Associations 173
imitation 28, 30, 36
impact
  on innovation 13–15, 57, 66, 77, 110, 112, 117–119, 132, 181
  of culture 13
independent inventor 17, 115, 161, 162–175, 177–184
indicators 13, 43, 53, 57, 62, 66, 67, 74, 75, 104, 108, 115, 119, 122, 125, 127, 135, 139, 141, 150, 192, 212
individualism 13, 59, 61, 63, 65, 66–70, 73, 76, 77, 94
Inglehart 10, 12, 20, 60, 78, 113, 125, 142, 149, 150, 156
Innovation
activity 10, 95, 103, 108, 112–115, 124, 162
agents 128, 165
dynamics 124, 187, 192
implementation 21, 64, 66, 67, 79, 167
Indicators 57, 62, 127–141
  models 17, 151, 185–187, 189, 191, 207
  networks 9, 17, 98, 157, 184, 186, 182, 195, 199, 200–202, 204
  performance 1, 3, 4, 13, 66, 106, 107, 108, 110, 115, 117, 121–124, 162, 184, 187, 190, 204, 212, 213
innovation system(s)
  national 94, 99, 110, 114, 115, 126, 162, 180, 181, 184
  regional 17, 97, 99, 103, 107, 115, 116, 120, 123, 124, 131, 132, 135, 136, 151, 152, 161, 162, 164, 165, 171, 173, 178, 180–182, 187, 205, 207
innovative milieu 3, 15, 81, 95, 97, 106, 123, 124, 151, 187
innovativeness 1, 4, 10, 13, 14, 25, 26, 18, 31, 32, 34, 37, 106, 108, 212, 213
interactive approach 129
inter-relation 187
intervening variable 11, 120
interview(s) 10, 17, 114, 115, 119, 165–176, 192, 193, 199
intra-academic 128
invention(s) 5, 12, 26–29, 31–36, 38, 39, 63, 64, 105, 106, 129, 161–163, 167, 169, 170–174, 178–182, 212
inventor(s) 4, 17, 29, 115, 161–184, 213, 215
Irawati, D. 16, 142, 143, 145–147, 156, 157, 188, 206, 212, 213, 215
Ivančič, A. 17, 161, 213, 215

J

Java 16, 143, 145–148, 154, 206
Jimenez, J. 48, 52, 55
John, R. R. 27, 30, 38, 79
Johnson, C. M 86, 99
Johnston, R. 26, 38
Joy, B. 45, 55

K

Kaasa, A. 13, 20, 65, 66–68, 76, 79, 112, 114, 119, 124, 164, 170, 183
Karlsson, C. 21, 180, 181, 187, 206
Kass, L. R. 42, 45, 55
knowing in action 90, 92, 95, 97
knowledge
  economy 3, 5, 20, 21, 25, 96, 98, 127–129, 131, 133, 141, 142, 148, 182, 184, 192, 205
  generation 80, 81, 89, 94
  Production 6, 19, 31, 48, 50, 55, 128, 129, 140, 153, 168, 205
  Society 21, 42, 140
  Transfer 3, 4, 6, 80, 90, 94–96, 103, 108, 119, 130, 135, 146, 147, 185
knowledgeability 14, 27, 32–34
Koepsell, D. R. 53, 55
Kurzweil, R. 40, 42, 45, 55

L

Lagendijk A. 145, 148, 157, 185, 187–190, 206
Landes, D. S. 35, 38
Larson 33, 38, 145, 157
Latour, B. 27, 38
learning regions 187, 189, 190, 205, 206
linear approach 129
local territory 17, 186, 202
log-linear econometric model 153, 157, 161, 182, 184, 206
lone inventor 161, 182, 184, 153, 157, 206
Lorenzen, A. 147, 149, 152
Luhmann, N. 35, 39

Index

Lundvall, B-Å. 3, 20, 81, 99, 120, 127, 141, 180, 183
Lynch, Z. 44, 55
Lyotard, J. 33, 39

M

Mannheim, K. 37, 39, 215, 210
Marx, K. 34, 39
Masciarelli, F. 8, 9, 19, 20, 124, 163, 167, 178, 183
Masculinity 13, 61, 65–71, 73, 74, 76, 77, 174
Maskell, P. 81, 88, 95, 97, 127, 141
Mast, J. 2, 14, 25, 216
Maurer, S. M., 46, 55
Measure(s) 33, 49, 50, 62, 63, 71, 109, 111, 112, 117–122, 171, 188
Mejlgaard, N. 48, 55
meta-analysis 15, 103, 114–116, 125, 140, 150
metaphor 14, 27, 29, 31, 60
Mitcham, C. 41, 46, 49, 55
Mode 1 129
Mode 2 129
Moldaschl, M. 1, 18, 20, 28, 38, 39
Morgan, K. 98, 142, 145, 147–153, 157, 189, 190, 207
Moulaert, F. 147, 152, 153, 157, 185, 187, 189, 204, 207
Mulhall, D. 44, 55

N

nanotechnology 42, 43, 45, 55, 56
national Ethics Committees 48, 49
Neira, E. 14, 57, 58, 79, 216
Nelson, R.R. 37, 39, 81, 99, 141, 182, 183
Netherlands 109, 140, 191, 93, 94, 206, 216
network(s)
  structure 9, 145, 200
  analysis 10, 115, 124, 203, 206, 213
neurotechnology 44, 47, 50, 53, 55
Nordmann, A. 47, 55
norms and values 4, 7, 18, 53, 82, 128, 149–151, 211
NUTS1 113
NUTS2 113, 120

O

O'Sullivan, D. 87, 100
OECD 53, 55, 62, 122, 127, 141, 162, 180, 184, 192, 204

P

Parsons, T. 11, 12, 20, 21, 35, 39
participation 9, 17, 32, 47, 48, 49, 52, 55, 84, 88, 89, 94, 99, 105, 107, 112, 114, 117, 119, 120, 151, 163, 178, 179
patent 1, 3, 13, 41, 43, 44, 53, 54, 55, 62–67, 112,117–120, 133, 150, 161, 163, 166, 170–173, 175, 178, 180, 181, 183
Pavitt, K. 37, 39, 130, 138, 141
Paxton, P. 162, 163, 184
performative(s) 29, 30, 31, 32, 185
Podmenik, D. 17, 161, 173, 181, 184, 216
Porter, M. 3, 19, 81, 100, 142, 143, 148, 149, 157
Portugal 13, 15, 49, 57, 67–70, 74–77, 215
Positivist 13, 113, 116
post-empiricist 116
post-industrialist 60
post-industrial 188
post-positivist 122, 123
professional communities 83, 93
prototypes 129, 170
Putnam, R.D. 7, 10, 20, 41, 52, 55, 81, 100, 107, 110, 121, 141, 163, 164, 184

R

R&D 18, 62, 63, 64, 67–74, 77–79, 103, 111, 115, 117, 118–120, 135, 137, 150, 162–164, 192, 211
R&D expenditure 64, 67–74, 188, 211
regional
  business environment 148
  development 20, 98, 112, 123, 124, 140, 141, 148, 151, 152, 157, 158, 182, 191, 192, 198, 204, 205, 206
  framework 108
regulation(s) 33, 37, 43, 48, 55, 76, 111, 138, 180, 267
relational proximity 88, 91, 92
responsible research and innovation 47, 56, 52
Roberts, J. 80, 81, 82, 83, 86, 89, 91, 92–100, 204, 213, 216
Roco, M.C. 40–42, 47, 51, 54–56
Rosanvallon, P.37, 39
Rosenberg, N. 37, 39
Routinisation 104, 116, 201

Rutten, R. 16, 17, 108, 112, 115, 119, 122, 125, 140–142, 148–150, 157, 185, 188, 189, 206, 207, 212, 213, 221

S

S&T production process 14, 41, 42, 15, 46, 48–53
Salamon, L. M. 37, 39
Sampling 9, 13, 108–111, 113, 115, 117–120, 212
Sartori, G. 27, 39
Saussure, F. 29, 39
Schieman, S. 26, 39
Schmidt, M. 40, 47, 51, 56
Schön, D. A. 31, 39
Schumpeter, J. A. 3, 4, 8, 21, 27, 35, 39, 132
Schutz, A. 49, 50
Scott, A. 2, 20, 140, 141, 142, 148, 158
Snow, C. P. 47, 56, 80, 99
social
  habits 17, 186, 202
  network(s) 4, 5, 9, 16, 28, 34, 52, 53, 83, 104, 108, 112, 119, 136, 137, 139, 140 151, 163, 175, 177, 186, 189, 190, 201, 203, 205
  relations 5–7, 16, 41, 106, 127, 128, 136, 138, 151, 164, 180
  relationships 44, 135
social capital
  Bridging 4, 5, 46, 47, 48, 51, 53, 93, 95, 106, 121, 142, 144, 451, 154
  Bonding 5, 47, 51, 93, 144, 212
  Regional 7, 9, 14, 132, 151, 153, 154, 155,
  Individual 163, 167, 17
  Structural 163, 167, 178, 179
socially responsible science and innovation 14, 40, 41, 42, 44, 47, 49, 51–53
Sombart, W. 25, 26, 35, 39
Spain 13, 15, 57, 67–70, 74–77, 216
spatial proximity 149, 152
speech act theory 30
Sprague, J. 37, 39
Stehr, N. 1, 2, 14, 18, 20, 25, 35, 38, 39, 44, 56, 216
supportive environment for innovation 171, 172
Swan, J. 86, 93, 100
Swidler, A. 11, 21, 29, 39

T

tacit knowledge 2, 8, 88, 128, 129, 130, 149, 152, 166, 178, 187, 190, 199
territorial innovation models (TIMs) 17, 18, 151–153, 155, 186, 187–198, 190, 198, 199, 200, 202
Torre, A. 142, 143, 149, 156, 158
Toyota 82, 146, 147, 156, 158
transfer 3, 4, 6, 80, 85, 86, 90, 93–96, 103, 108, 127, 189, 130
triple-helix 2, 19, 140, 126, 134
Twente 185, 191–207, 215
Twente Technology Circle THT 194–196

U

uncertainty avoidance index (UAI) 12, 13, 61, 64, 65, 68, 70–73, 169, 170, 181
University of Twente 196, 204, 206, 207

V

values 12–13, 67, 75, 77, 78, 82, 93, 118–122, 125, 128, 134, 139, 142, 144, 147, 149, 150–152, 154, 156, 157, 80, 207, 211, 212, 213
Van Lieshout, M. 50, 56
Vieira, I. 14, 58, 216
virtual communities 92
Von Hippel, E. 94, 100, 169, 184, 37, 39
Vromen, J. J. 28, 39

W

Waclawski, J. 36, 38
Watson-Verran, H. 36, 39
well-informed citizen 49, 52, 56
Wenger, E. 80, 81, 84, 85, 57–89, 92, 95, 96, 100, 185, 208
Westlund, H. 1, 5, 6, 16, 21, 115, 121, 125, 126, 130, 133, 141, 158, 162, 164, 179, 184, 208, 211, 216
Williams, E. A. 13, 21, 115, 121, 125, 126, 127, 130, 133, 141, 158, 162, 164, 179
Wolfe, D. 132, 140, 141, 164, 184, 188, 205
Woolgar, S. 28, 39